A Very Different Land

A Very Different Land

Memories of Empire from the Farmlands of Kenya

Hilary Sunman

The Radcliffe Press
LONDON • NEW YORK
An imprint of I.B.Tauris

Published in 2014 by The Radcliffe Press
An imprint of I.B.Tauris & Co Ltd
6 Salem Road, London W2 4BU
175 Fifth Avenue, New York NY 10010
www.ibtauris.com

ISBN: 978 1 78076 996 7
eISBN: 978 0 85773 681 9

A full CIP record for this book is available from the British Library
A full CIP record is available from the Library of Congress
Library of Congress Catalog Card Number: available

Typeset in Sabon by Saxon Graphics Ltd, Derby

For my parents and their
generation of colonial servants

CONTENTS

LIST OF PLATES

FOREWORD

As a teacher of history, it is difficult to explain the British Empire to today's students. Its geographical expression at any one time is easy enough. It is the range of imperial motivations that is tough – the spectrum that stretched from the monstrous to the altruistic with pretty well everything in between. Empire as a phenomenon lays itself open to parody, disapproval veering on condemnation, conspiracy theory and monocausal explanation. That is why books like Hilary Sunman's are so important. The richness, texture and carefully sensitive recreations within these pages are a powerful and necessary antidote to those who come to imperial history to sneer and smear bearing their broad brushes and primary colours.

Hilary loved the two central characters of the book. But the world of Owen and Dorothy draws in so many others. She has a deft touch for character, mood and moment. Lytton Strachey, in his fabled life of Queen Victoria warned biographers of the perils that await those authors who seek to penetrate 'the secret chambers of consciousness'. But Hilary is very good on motivations both of Owen and Dorothy's generation and that of her own – those who had a late imperial childhood, became the first post-imperial generation, sought to come to terms with it and did so by harnessing those altruistic impulses to the cause of aid and development. Her 'Reflections – children of empire' chapter is fascinating and brimming with both a personal and a wider perspective.

What really adds to the fascination of *A Very Different Land* is the Sunman family bond. It is also a treat to read and absorb. It passes, too, the great test of any book – it leaves you wanting more. How intriguing it would be if Hilary turned sociologist and social anthropologist by comparing her fellow professionals, who did their training just as the British government completed its dash for the imperial exit, and those of Owen's who completed their training when the British domination of palm and pine seemed set fair for many decades more than turned out to be the case. A sensitive analytical mind and a fluent pen are needed to rescue Owen and Dorothy's generation from what Edward Thompson called (albeit in very different historical context) 'the enormous condescension of posterity'. These Hilary possesses, and as this book shows she is as at home with high policy making as she is with Owen trying to find ways of protecting the maize from weevils in Nyanza Province.

Few in Hilary's and my generation are neutral about empire. It still has the capacity to induce an emotional spasm – sometimes several conflicting spasms given the range of imperial motivations and experiences. We have a desire to make careful historical sense of it and to distil its better essences for the purposes of aid policy and development funding today. Hilary's book is a very considerable contribution to both.

Peter Hennessy, FBA
Attlee Professor of Contemporary British History
Queen Mary, University of London
April 2013

ACKNOWLEDGEMENTS

In the writing of this book I have been fortunate to meet a number of people who have inspired and helped me, in Kenya and in England. An excellent starting point was Peter Ayre and his wonderful database of all Europeans who lived and worked in Kenya between the wars. He introduced me to James Foster and Roy Griffin in Nairobi, both closely involved in the cathedral and the choir. James shared material from his biography of Leonard Beecher, while Roy, who sadly died recently, offered thoughtful memories and anecdotes, giving me a great sense of continuity from the period I am writing about to the present. Roy introduced me to Rosemary Gardner and her wit and charm and encyclopaedic address book put me in touch with many who have been involved in the musical life of Kenya in the past 50 years. The Rhodes House collection of the Bodleian Library in Oxford has been a rich source of archive material and documents on Kenya and agriculture of the period. In Kenya, Titus Murithi from Meru, Lukas Alebe and Washington Omundi were charming and helpful. Bridget Cass introduced me to her great friend Elizabeth (June) Watkins, who welcomed me to the 'elite band of writers about Africa' and set me right on many things. Kate Barker helped me reshape the book to bring more life to the characters, which is as well, since I must also thank Peter 'you must write about these people' Hennessy without whom I would probably not have embarked on the project. Finally, Peter, my husband; I have often mused on the effusiveness with which writers thank their spouses, but now I understand it. He has read every word – more than once; he cooked for me; sent me off to purdah to concentrate on the writing; and has supported me and the project throughout. Thank you, Peter.

LIST OF PEOPLE

This story contains references to a number of characters, all real, starting with my parents, Dorothy and Owen Sunman. The principal others are listed below.

Agricolas and administrators

Denzil Blunt
John and Stella Booth
Ted Gaddum
Stuart Gillett
Alex Holm
Tom and Margaret Hughes Rice
Roger Lambert
Bill Leckie
Bill and Hilda Lyne Watt
Cedric Oates
Donald and Betty Storrs-Fox
Tom and Margaret Watson

Missionaries

Dr John Arthur
Leonard Beecher
Canon Henry Leakey
Archdeacon Owen

Musicians

Philip Angore
Valerie and Tony Davis
John Dixon
Bronslaw Fryling
Rosemary Gardner
Roy Griffin
Graham Hyslop
Billy Isherwood
Arthur Kemole
Nat Kofsky
Atigula Luvai
Peter Mence
Dick Moss
Ben Ogola
Francis Olude
Washington Omundi

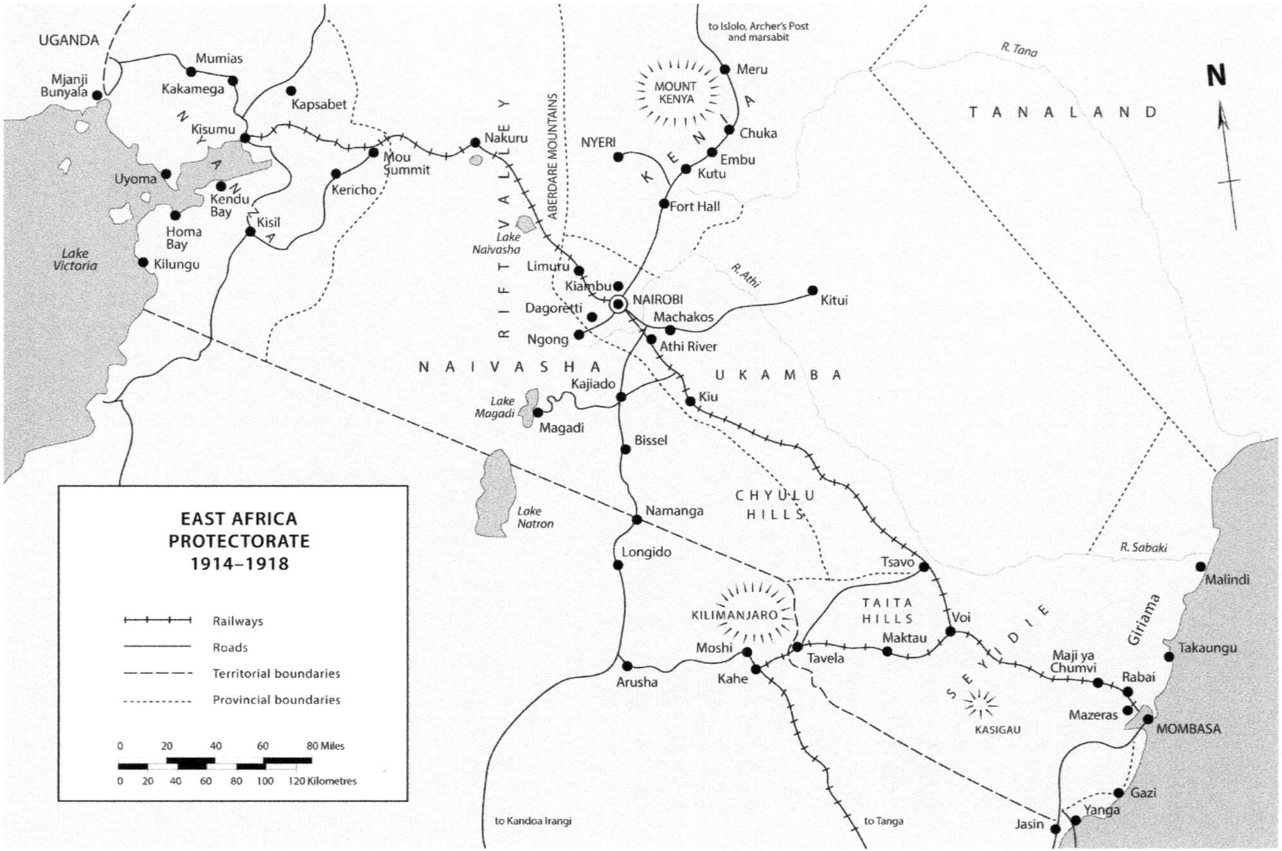
EAST AFRICA PROTECTORATE 1914–1918
Railways
Roads
Territorial boundaries
Provincial boundaries
0 20 40 60 80 Miles
0 20 40 60 80 100 120 Kilometres
N
UGANDA
Mjanji Bunyala
Mumias
Kakamega
Kapsabet
Kisumu
NYANZA
Uyoma
Kendu Bay
Homa Bay
Kisil
Kilungu
Lake Victoria
Kericho
Mou Summit
Nakuru
RIFT VALLEY
ABERDARE MOUNTAINS
Lake Naivasha
Limuru
Kiambu
Dagoretti
Ngong
NAIROBI
Machakos
Athi River
NAIVASHA
Kajiado
Lake Magadi
Magadi
Bissel
Lake Natron
Namanga
Longido
Arusha
Moshi
Kahe
KILIMANJARO
Tavela
Maktau
TAITA HILLS
Voi
Tsavo
CHYULU HILLS
Kiu
UKAMBA
Kitui
R. Athi
NYERI
Fort Hall
Kutu
Embu
Chuka
Meru
MOUNT KENYA
KENIA
to Islolo, Archer's Post and marsabit
R. Tana
TANALAND
R. Sabaki
Malindi
Giriama
Takaungu
SEYIDIE
Maji ya Chumvi
Rabai
Mazeras
MOMBASA
KASIGAU
Gazi
Yanga
Jasin
to Tanga
to Kandoa Irangi

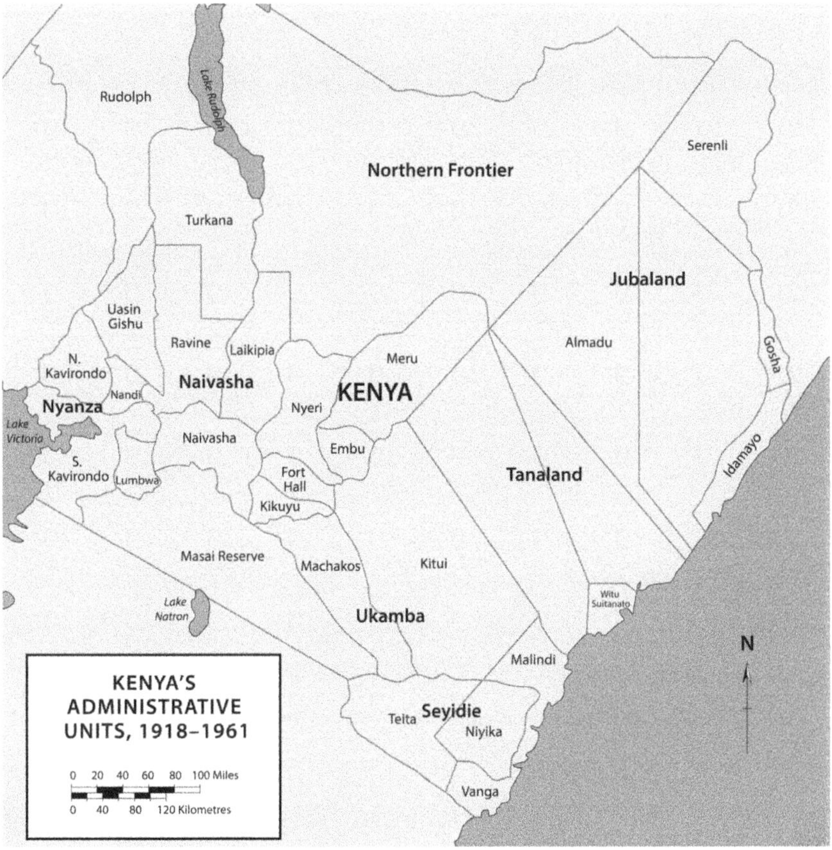

KENYA'S ADMINISTRATIVE UNITS, 1918–1961
0 20 40 60 80 100 Miles
0 40 80 120 Kilometres
N
Rudolph
Lake Rudolph
Northern Frontier
Serenli
Turkana
Jubaland
Uasin Gishu
Ravine
Laikipia
Meru
Almadu
Gosha
N. Kavirondo
Nandi
Naivasha
KENYA
Nyanza
Nyeri
Lake Victoria
Naivasha
Embu
Idamayo
S. Kavirondo
Lumbwa
Fort Hall
Tanaland
Kikuyu
Masai Reserve
Machakos
Kitui
Lake Natron
Witu Sultanato
Ukamba
Malindi
Teita
Seyidie
Niyika
Vanga

INTRODUCTION

'Be careful what you wish for'

Mention colonial Kenya, and people will think of two things: the notorious Happy Valley set (sundowners from sundown to sunrise, darling) and Mau Mau – and will generally conclude that the antics of the former led directly, inevitably, to the latter. But Kenya was always part of my life and it is difficult to see one's background in such stark and one-dimensional terms.

I was born in Nairobi in 1947 and I grew up in England surrounded by talk of life in Kenya, and it was always a benchmark of comparison and memory. My parents both loved their life there and my childhood was full of reminiscences and Swahili phrases; when we went to France and my parents were searching for a word in French, out came a word in Swahili. A stool made from a single piece of wood and decorated with beads came from Kisii; a round coffee table in an African hardwood was bought in Kitale. When the weather in England got very hot my father sighed with happiness and said it reminded him of Mombasa. When, later, my mother visited me in Hong Kong, she loved the verandahed Repulse Bay Hotel because it reminded her of the Mombasa Club.

Owen, my father, joined the Agriculture Department of the Colonial Service in Kenya in 1928 at the age of 23. He rose from lowly assistant agricultural officer to assistant director of agriculture in 1948, working mainly in African agriculture. After six years, in 1934, he took home leave and returned with Dorothy, his wife, and together they lived in Kenya for 15 years, until Owen retired from the Colonial Service at the age of 46 to take up a life in England. My mother brought me back to England in 1949, in a flying boat from Lake Victoria. He followed in 1950.

My life has echoed my father's. I have worked in development and technical assistance as an economist in developing countries for 30 years, sometimes in Africa but also Asia and the Far East, and the Middle East – although never Kenya. I had not realised how close the parallels are between his world in Kenya in the 1930s and 1940s, and my development world in the latter part of the twentieth century, until I started reading some of the reports by him and other agricultural officers in the departmental reports and recognised the language of the adviser; there is a characteristic tone of weary optimism. 'This land is very fertile and we hope that with sufficient demonstration and teaching the local population will be able to increase the output of maize and food crops…' or 'Once the road has been completed costs of transport will be much reduced which will enable better marketing of local produce. The benefits to the local population in terms of nutrition and household income will be significant'. A willing for things to improve but a recognition of the great difficulties; 'if only things were organised like this, or like that…'.

When I began to write this book, I had two main thoughts: to explore what Owen and his colleagues thought they were doing, why they had come and what they achieved; and to examine what is left of their time in Kenya – and what they contributed to the life of the country. I wanted to look at the expectations and background of my father and those like him who joined the Colonial Service, what they hoped for, how they had found out about it. Was it to be heroes and adventurers? Missionaries and teachers? Or was it to build a sound administration? What was the life like, day to day? What was it like for the wives and women who joined them? Mine is the last generation with direct connection to the colonial officers of the first half of the twentieth century, and there seemed to be a story to tell. I wanted to look at their lives and their experiences, knowing that at the time when Owen became part of the Colonial Service, it was seen as that – as service. And, at the time, 20 years in the Colonial Service was seen as a good basis for a future career 'back home'.

The history I am telling, of the lives and aims and achievements of a generation of technical advisers, colonial servants, is undoubtedly a contested one. Now, 50 years after Independence, there is some recognition within Kenyan society that not everything that the British did was bad – 'we must have done something right'. But in post-Independence Kenya this would have been a hard message to put across. And I, who have worked as a development economist for 30 years, shared many of those post-colonial perceptions moulded by a Marxian analysis of the objective truth of imperialism, not to mention a heavy share of post-imperialist guilt. Writings of African writers – N'gugi wa Thiongo of Kenya, Chinua Achebe of Ghana, Dominic Mulaisho of Zambia – left me in no doubt as to the corrosive and damaging impacts of empire and her colonial servants. Indeed, while I was studying for my Master's degree in Development Economics, I told my mother (my father had died by then) that she and her peers were simply exploiting the Kenyan people, an accusation that upset her considerably. So perhaps my goals in working in development were at least in part to redress some of this balance; and perhaps they stemmed from a yearning to travel and discover the exotic of the world – a drive surely shared by the young colonial servants who had joined the service 40 years before.

In this book there are a number of themes. This was a period, in Kenya, where government allocation and gazetting of land was complete, and white settlers were well established and farming with a range of skill, success and commitment. The London administration saw itself as building technical and governance capacity in the colonies, with a view to future self-government and perhaps independence – possibly two or three generations hence. On the surface, the 20 years of Owen's service was a period of stability and administrative ordinariness and was seen by the colonial government as a period of consolidation and progress towards a better future state for the colony as well as for the British government. By the end of the 1920s, the colony was poised between the early 'Wild West' settler period and the Second World War, a period which has been called 'the golden age of the colonial administration'. It was also a period of rapid change and crises, particularly the world depression. At the same time education and mission schools began to play an important role in changing African perceptions and capability. The development of agricultural policy was crucial to Kenya's future, to the development of the African population and to the life of my father and others like him. But his career was tempered by tensions and hostilities that accompanied the early years of Kenya – tussles between settler and government, between settler and the Indian population and between settler and African – all of which centred around land, which lay at the heart of politics in Kenya.

Many of those joining the Colonial Service would have been idealistic young men seeking to help to build and develop a new society. But events and perhaps their own actions hastened the creation of a world very different from the one they envisaged and they would have been horrified had they foreseen the grave tensions and anger that led to freedom movements and ultimately the Mau Mau uprising.

So I have tried to understand more of the actual life of a colonial administrator, particularly a technical adviser. Life involved the technical aspects of agricultural extension work and training, and later as he and his contemporaries matured, development of policies for the sector. What was the life and how did they fit in with the rest of the Kenyan community? It was clearly a good life and involved music, the church, sport and bridge, social life and mutual support, all of which have left tangible legacies. Yet it must have been a strange one too; cut off from the day to day life of Europe. By the time of the Second World War, colonials were seen by many back home as tangential and war dodgers, despite the protected nature of their work, especially in agriculture.

What were the legacies of this period? Most of Owen's career was in the Department for Native Agriculture, and this has allowed me to reflect on how much support and effort went into strengthening African, as opposed to settler agriculture, whether enough was done and if not why not. But the agriculture developed by the European and African farmers has become the backbone of Kenya's agricultural sector today, and the Department of Agriculture, the officers with their experimentation and demonstrations, was crucial to the lasting success of coffee, tea, maize, flowers – all the crops that are produced today. There is a lasting agricultural legacy.

There were other legacies too. Owen became very involved in church music, at the cathedrals of both Mombasa and Nairobi, and there is a living link through music to the time when he lived there. I did not return to Kenya until 1980 for a one night stopover between Lusaka and London, but when I did, I visited All Saints Cathedral in Nairobi and found my father's photograph in the vestry; he had been choir master from 1945 to 1950. One of his close friends, a teacher of music, wrote the Kenyan national anthem at Independence, at the request of Jomo Kenyatta.

Meanwhile in the rest of the world and London, the colonial venture was seen on the one hand as of enduring strength and on the other as something to be shed as quickly as reasonably possible. The establishment of colonies and empire had always been controversial – Gladstone wanted nothing to do with it – and in the forties and fifties, particularly following the war, the whole programme was increasingly irrelevant to many. Within Kenya there was a growing caste of educated Kenyans – and a growing articulation of desire for change, for independence. Popular opinion in the UK and Europe saw the colonies as increasingly anachronistic, speaking of a different age. Twenty years of colonial service, rather than being an asset, became seen as a hindrance in the post-war British job market.

It is impossible to write of Kenya in the late 1940s without referring to the Mau Mau. It is still a highly contested story and to write anything is to fall out with some set of ideas. So I had been hoping to avoid it. I am looking at a specific period – 1928 to 1950, when my father was in Kenya, and I am focusing on native agriculture and other possible legacies from this period, and the Emergency was not declared until 1952. But the politics of Mau Mau directly affected the lives and work of agricultural officers during the 1940s, years before the declaration of the Emergency.

But the book also became for me a journey of understanding. I have understood that the legacies are many and tangible, and far from all negative. The structure of agriculture in Kenya – a sector which today accounts for about one quarter of GDP – builds directly on the labours of the early settlers and the Department of Agriculture. Church music is strong throughout Kenya – although perhaps not in the classical model – and there is, increasingly, recognition of the contribution of dispassionate governance during the colonial period. The changes in Kenya in three generations since Independence are profound – yet there are links and continuity which can be traced to those earlier days. Not all outcomes were expected; many plans were not fulfilled – many outcomes were unexpected – 'be careful what you wish for'.

There are wider legacies of empire too. There is the sense that the British continue to hold on to being bigger players on the world stage than they can realistically justify nowadays. This is perhaps exemplified in the dominant role of British consultants (my career) in the world and perhaps in the enthusiasm for intervention in tricky global situations – a sense of a special British quality which may underlie why so many citizens do not wish to be part of a community, the EU, where the UK is simply one of 27 countries, the majority of which are much smaller than the UK, and do not share the same imperial history. It also explains the continuing constant rivalry between France and Britain, the only two members of the EU to have had a significant empire.

This is a wider picture, but the main aim of this book is to explore the 1930s and 1940s, seen through the prism of the lives of colonial servants and their colleagues. The outcomes for this generation were very different from what might have been imagined in the 1920s, yet they have made a real contribution to modern Kenya.

In writing this book I have used interviews and memoirs, as well as government records. I have visited Kenya several times, and have been to all the places where Owen and Dorothy were posted: Meru, Mombasa, Kakamega and others, and talked to people in each. Libraries and archives have provided great riches, from the names of vessels steaming to Africa in the 1920s to detailed reports by the Department of Agriculture for every year from 1921 onwards. While few of his contemporaries are still alive (he was born in 1904), there are people who knew of him, and I have followed up with their children – my contemporaries – where possible.

I was fortunate enough to find a number of memoirs of officers from that period in Kenya in the Rhodes House library collection in Oxford; for the details on agricultural practice and the lives of agricultural officers, I found unpublished manuscripts which were the background for a review by Oxford University on agricultural practice in Kenya, part of the Oxford Development Records Project undertaken in the early 1980s under the direction of Anthony Kirk-Greene from St Anthony's College, Oxford. There are one or two extraordinarily rich databases which document the lives and people of that period – that of Peter Ayre will be well known to followers of Kenya. He is compiling a database of all the Europeans living in Kenya between the wars, and this record has been extremely valuable. Tim Hutchinson has another extremely useful directory of most of the Europeans who lived and worked in up-country Kenya.

A word about language and terminology is needed here. Nowadays the term 'native' is seen broadly as derogatory, but in the 1920s it was the language of the government and embodied in many terms: the Native Affairs Department, the Department for Native Agriculture and so on. Where it seems appropriate – mainly in reference to official documents and policies – I have continued with this usage but also use the terms African and indigenous peoples. During the 20 years I am looking at the use of language changed within the government too. The early official reports spoke of 'native farmers', but gradually the term 'planters' comes to replace this. It must also be remembered that the terms 'nigger', 'fuzzy wuzzy' and 'wog' were in common usage, so that native is a relatively polite term. The Agricolas about whom I am writing were in the main determined for the betterment of native agriculture and quality of life, a perspective which probably outweighs the use of the term 'native'.

Part I

Calls of Empire

1

EXPECTATIONS AND ARRIVAL

'To be an Englishman was to be born to a great destiny; as warrior and guardian of freedom and justice and peace... he had already decided to be a missionary or a colonial officer, he could not decide which.'

Joyce Cary, *A House of Children*, 1941

'They saw themselves as missionaries of science bringing hope and prospects of progress to underprivileged peoples in an undeniably beneficial task.'

Masefield, *A History of the Colonial Agricultural Service*, 1972

On 17 February 1928, at the age of 23, Owen Sunman alighted at Kilindini Harbour in Mombasa after a five-week voyage. When he left London the weather was wintry – typical January weather: squally and penetrating south-westerly winds, wintry showers and frost at night. The journey had taken him through the Mediterranean, still wintry, and then to Suez and through the Gulf of Suez. By Port Sudan and Aden the temperatures were almost too hot and the boat was unbearable – but on arrival in Mombasa with mellow temperatures of around 80°F, gently wafting palm trees and the glittering white buildings of the old Arab port, the chill of England was already a lifetime away.

The journey began at Victoria station, on 28 January – his sister's birthday – with the boat train to Paris and on to Marseilles where he was to board the SS *Bernadin de St Pierre* bound for Mauritius. The station was a hubbub of friends and family, porters, baggage, last-minute purchases, farewells, tears and laughter. Passengers were warned to arrive at least three-quarters of an hour before departure at Victoria because of the crowds. Family and friends came to see off the departing; this was a ritual of wellbeing which would be replicated again in Nairobi. Owen's mother, father and sister were there, wrapped against the bleak cold and perhaps the sadness of seeing their beloved boy sail off to who knows what. They would not see Owen again for three years when he returned for his first long leave. Letters and cables were reliable but slow. Jossie, his older sister adored him and missed him deeply; it was a poignant coincidence that the date of sailing fell on her birthday. His friends from Clock House church, Eric and Maurice Ashby joined the party, together with family friends and neighbours, Charles and Ethel Adams and their two girls, Dorothy and Joan, to share the excitement. At 23, Owen was musical, good looking, clever – a first-class degree in Agriculture from Wye College under his belt, months spent working on an experimental farm in East Anglia and a spell on the wheat steppes of Canada. He was full of confidence, exhilaration at the freedom ahead, the vocation of colonial service, the briefings and instructions for the journey ahead. For his family it was a more bittersweet occasion, a mix of pride and anguish – but his friends may well have been envious of the voyage ahead.

All luggage had been sent the day before – two tin trunks seemed to be the normal allowance. In addition, handbooks for the colonies listed essential items: water filters, a kettle for boiling water, a sun helmet (since 'exposure to the sun should be avoided at all times, even when the sky is cloudy and the temperature appears cool. A sun helmet should invariably be worn when out of doors; a cap or straw hat only after sunset'). Only a few years earlier it was considered essential to wear underclothes of red flannel, to avert the impact of dangerous actinic rays from the sun.

Although the houses for colonial officers were now furnished with the bulkier items of furniture – this had not always been the case – there was still a hefty list of equipment to pack and take.

In addition to a travelling bath, cutlery and crockery, candlesticks and lamps, cooking utensils ('those of aluminium in "nest" form are convenient'), a range of items was suggested in the guidelines: a couple of easy chairs or folding chairs with cushions; a rug (available at the Army and Navy Store, in Victoria Street in London); light Indian curtains and table covers. Clothing included tennis shirts with soft collars; grey or white flannel trousers; two or three suits as well as evening dress clothes with white shirts; khaki and white drill suits; a good supply of boots; puttees; leather gaiters; a great coat; a travelling rug; and a waterproof with sewed seams. And tennis shoes, a racket, a hockey stick, golf clubs and balls and a cricket bat. On ocean liners an officer was permitted 80 cubic feet or 2 tons of baggage free, and half that on the railway. It is not clear what happened once the officer arrived at Mombasa and had to reduce his luggage to comply with the limitations of the train. Perhaps the surplus followed after him over land on slow drawn ox cart.

Once at sea – after a night in Marseille and Owen's first taste of a perfect French omelette – life on board was measured out by meals. Breakfast at 7.30 sharp, and if you were after 8 there would be no eggs; lunch at 10.30, tea at 4.30 and dinner at 6.30; and rubbers of bridge. Junior officers were berthed in modest accommodation where four men were crammed into cabins better suited to two – with no chest of drawers, just a rack overhead for clothes. The social status of those on board sorted itself out: the colonial officers separate from the settlers and among the colonial servants a sharp distinction between the higher echelons – the Brahmins – and the cadets, who might be favoured with a 'good morning' but little more from their superiors. New recruits – like Owen – were lowly creatures compared with the senior administration but the colours of sunsets over the desert, the green of the phosphorescent sea, and the effervescent social life made the voyage enchanting. On every voyage there was an enthusiastic, organising passenger who arranged games and tournaments, energetic sports like deck quoits, skittles and deck tennis, and more sedentary ones like cribbage. Bridge filled hours and there was Tombola after dinner.

Owen was sporty and played tennis and quoits and, always, bridge. There was a cinema on board; there was dancing every night and Owen doubtless brought his bassoon out and 'jammed' with the band who were described as 'painstakingly good'. 'Crossing the line' was an excuse for a party and the last night was often a fancy dress dance and a variety performance. By the time they arrived in Mombasa friendships had been formed, half romances tasted.

The journey itself was an adventure, and there were numerous calls in ports for refuelling and an opportunity for some tourism as passengers could alight at some of the ports. The liner passed from the green and blue of the Mediterranean through the hectic noisy Egyptian ports of Port Said, Ismailia and Suez, desert landscapes lining the great Suez Canal, with road and rail running alongside the canal in the empty desert, and then down to the burning heat of Port Sudan and Djibouti beyond. Port Said was truly the last link with Europe for among the souks and the hustling from local traders was a department store called Simon Artz, owned by Selfridges of London, where the young colonial servant could equip himself with a pith or cork helmet (25 shillings for a solar topee), tropical clothes and anything else that he might have forgotten to get from the Army and Navy stores. Colonials on their way home also bought clothes there to replace wide brimmed felt hats or topees – a trilby hat, more suited to England.

The ships took on coal at Suez and Port Said, a very noisy business as 500 tons of coal, carried in baskets, were pitched from men's shoulders into the bunkers on the ship. There was a nice bathing place at Ismailia, where the canal was guarded by great stone lions on the memorial from the First World War. In summer the Red Sea was unbelievably hot, but in February would have been a delight after the chill of the English winter, a first taste of African climes. The crew switched their blue uniforms to white tropical kit, winter sweaters were shed. At Port Sudan the thing to do was to take a glass bottomed boat and look at the fish, or to take a farouche and drive around the town with its garden, canal and camels. Djibouti was described by one traveller as 'a most dreadful place full of dirty Arabs [sic], flies and every sort of beastliness, hot as Hades, not a tree or any vegetation'. But Arab boys entertained, crowding around the ship, diving for coins with extraordinary energy; they could dive as far as the coins were flung, and collect them and sing while holding them in their mouths.

Fellow passengers were varied in their means and purposes: nurses, mission folk, priests, settlers, as well as colonial officers. It was a French ship so there were French monks going to Mauritius and French nuns for the Belgian Congo (an easier journey for the Eastern Congo than sailing round the great bulge of West Africa). There was a Czech settler who was a champion big game hunter. Many travellers took advantage of the time on board to get to grips with Swahili, the most widely spoken language in Kenya, using *Swahili Exercises* by Steere available from the SPCK[1] bookshops. Colonial officers were required to pass exams in Swahili and reach a decent standard, otherwise pay increments and promotion would be withheld. Owen teamed up for lessons with an old parson and a nurse who was a Swahili expert, and who were going on to Zanzibar to do mission work.

But despite the entertainment and social life, after five weeks of voyage crammed into a small cabin, arrival in Mombasa with its exotic smells and sounds was exhilarating and overwhelming: the green of the palm trees; the purple of the bougainvillea; the richness of women's clothes – gaudy colours of printed cottons worn by the Swahili women, the silken glow of Indian saris; the rich tones and hubbub of Mombasa were truly welcome. Rickshaws and narrow streets like Malta, open verandahs on the hotel and lizards provided a luscious contrast to the drab weather left behind in London, five weeks before. There was even a view of the Mombasa golf links as the boat dropped anchor in Kilindini Harbour. Passengers were rowed ashore and dropped on a wooden jetty to a large, shabby, wood and corrugated iron Customs and Baggage shed, where 2 tons of baggage were carefully checked.

Life on the Mombasa streets was noisy and mysterious: the crowds of Arabs on donkeys, heavily laden camels, rickshaws drawn at great speed by small African boys, women hidden behind shawls and jewels. This must have felt like the beginning of the adventure which brought Owen and others to the magic of Africa.

On this trip there were a few other colonial officers alighting at Mombasa, but no one else from the Agricultural Department. There was one other new arrival, an education inspector, and several other officers who were returning from leave: a medical officer, an assistant inspector of police, a registrar of land title; and two ladies, one a hospital matron and one a postal clerk. The new arrivals were met in Mombasa by the local district commissioner and then put on the train to Nairobi, when the timetable permitted. But first there was a night in a hotel, perhaps the Palace Hotel, standing alone opposite the French Mission, or the more central Manor Hotel, with open verandahs, lizards, cool beers and a bath – a place to stretch out after the cramped shared cabin which had been home for the past five weeks. The Manor Hotel was also the home of Liza, a giant land tortoise that roamed Mombasa Island in those days – she was said to be over 200 years old and was thought to have originated with Portuguese trading ships, as the Portuguese used to carry tortoises as fresh food. She wandered around the island, though based herself at the Manor. One day sadly she made her way over to the Palace Hotel, where she met her doom; she was resting near a taxi rank when a driver reversed sharply into her – her carapace broke the petrol tank but the tank also punctured her shell. She staggered back to the Manor and died a few weeks later.

The railway had started carrying passengers in 1900 but conditions were still primitive. There were not yet corridors or restaurant cars on the train (these were introduced in 1931) and progress up to Nairobi was slow with periodic stops for meals. The train left Mombasa around 4.30 in the afternoon, rumbling out across the plain to Samburu where Owen and the other first class passengers descended for dinner, which was served in a long smoky room set with trestles tables by Goan waiters in spotless white robes. The food was notoriously bad and, apparently, always the same: lentil soup, fish heavily disguised with tomato sauce in case it was slightly off, a tough beef or mutton stew and a lukewarm fruit salad with bright yellow custard. The carriages were cramped but while dinner was taken the boy could unwrap the bed roll and make up the bed. After dinner the train rumbled on for a brief stop at Voi by 11pm.

By morning the train reached 4,500 feet, onwards and up, an amazing journey on a great curving track, so that from the middle of the train you could see both ends of it in a great U shape, until it switched away and the curve lay on the other side of the train. All windows were tightly shut against the fine penetrating red dust from the Taru Desert. The Athi Plain rolled away for miles, thick with wildebeest, zebra, all kinds of deer and ostriches. The temperature dropped from the heady heat of Mombasa to a more familiar 65 or 70°F as the train crept up to 6,000 feet, and reached Nairobi. Breakfast was taken in Kiu, and the train arrived in Nairobi in time for lunch and the start of a new life. All journeys began and ended with meeting and greeting – a sort of party – and the platform at Nairobi would have been alive with friends and family – Africans, Indians and Europeans – welcoming the traveller to a new life – and there were many who were called, for a great variety of reasons, to try their luck in the colonies.

For in the early part of the twentieth century the colonies were a strong draw for young men in England. Part of the appeal was adventure – who would not be drawn by descriptions of the African dawn, of vast skies and animals roaming free, and the freedom to make a mark on remote and beautiful countries? But service was important too – either in the sense of the colonial administration, serving to help civilise and develop Africa, install a rule of law and the basis for modern life; or as a missionary bringing God to heathens, and at the same time education and medicine; or as a technical adviser or practitioner – the agricultural service, the medical departments. At this time, as

Margery Perham, later director of the Oxford Institute of Colonial Studies, notes,[2] people 'of my generation' (she was born in 1895) were taught from their school days 'that our Empire was a splendid achievement, conducted as much for the good of its many peoples as for our own... the words "trusteeship" and "partnership" held serious meaning'. Empire – with its cultural, political and economic relations – was valued highly, and when the BBC was founded in 1923, its first Director-General, John Reith, saw empire both as a source of broadcasting material and as a topic of central concern to national life; he had an almost mystical feeling for empire, which he regarded as the most successful example of internationalism and peaceful coexistence of modern times; the medium of radio could, he believed, contribute to the cohesion of British subjects and the worldwide family of English speaking peoples.[3]

Enthusiasm for empire was demonstrated in the great British Empire Exhibition held in Wembley in 1924–5. The British Empire was literally everywhere in the early twentieth century – at that time, the Empire contained 58 countries – a quarter of the world's population lived in countries painted imperial red on the map and it is of this period that it was said that the sun never set on the British Empire. The glow and pride that one can read from descriptions of the exhibition would have been a part of the consciousness of boys and girls growing up in the early part of the century. The shadow of the First World War was felt to have passed, and this great exhibition – costing £12 million (over £600 million today) and attracting 27 million visitors – expressed a sort of exhilaration and confidence. The official aim of the exhibition was 'to stimulate trade, strengthen bonds that bind mother Country to her sister states and daughters, to bring into closer contact the one with the other, to enable all who owe allegiance to the British flag to meet on common ground and learn to know each other'. The Kenya pavilion showed large photographs of great rolling plains and mountains, an immense variety of wildlife from antelopes to elephants, and African peoples, exotic with their spears, drums and humped, horned cattle – which would surely seize a young man's imagination.

It was an exhibition of sweeping vision, culminating in the grand Pageant of Empire, performed between 21 July and 30 August, with a cast of 15,000 people, 300 horses, a 100-piece orchestra and an enormous choir of local choral societies. The highest echelon of British composers brought works to it – Hubert Parry, Eric Coates, Granville Bantock, Arthur Sullivan and, of course, the newly appointed Master of the King's Musick, Sir Edward Elgar, who composed the Empire March and a series of songs with words by Alfred Noyes. George V's great speech from Wembley[4] was relayed on loudspeakers outside major department stores and the crowds were so large that they stopped the traffic. An estimated 10 million in total heard the King's words in 1924. Education was not ignored; the management of the exhibition asked the Imperial Studies Committee of the Royal Colonial Institute to assist them with educational aspects of the exhibition, which resulted in a 12-volume book *The British Empire: A Survey*, published in London in 1924.

Empire caught the minds of young people and students. If Owen went to the exhibition – as well he might have, living as he did in London and with a passion for music – he would have been caught up in the excitement and idealism. Working in the colonies would combine the idea of service with the notion of sharing in and building a grand global project.

He was not alone. The Colonial Office was stronger than ever and recruitment reached a peak of 497 people in 1928. Earlier the idea of the colonies was dominated by the swashbuckling early settlers, the pioneering but often ruthless early businessmen like Cecil Rhodes for example, but by the mid-1920s the idea of colonial service as a good career had caught the student mind and, among the professional classes in Britain, it was now thought to be as good an option as other more conventional routes such as law or the domestic civil service. It was attractive and competitive, combining the qualities of the British professional career with a sense of service – and great adventure. Yet at any one time the numbers in the Colonial Service were very small compared with the task in hand – the ambitions of development and civilisation. With reference to the 'thin red line' of the British forces at Balaclava, the Colonial Service has been referred to as a 'thin white line'.

An important change at that time was the increasing call for technical advice to support the administration. Administrators – the district and provincial officers – had hitherto formed the largest group within the Colonial Service, many of whom joined the service from a public school/Oxbridge background and with a degree in Classics. Kenya was typical in that, over the whole colonial period, nearly half of the administrative service – 293 out of a total of 623 – were from this background, which was thought to give a fitting intellectual framework for the efficient and largely dispassionate administration of large swathes of land and population, although it was one which equipped them less well for coping with and understanding the political changes which would emerge during the first half of the twentieth century. But that was all to come.

Still, the mid-1920s was a time of changing attitudes, and a new philosophy of the colonies, reflected a growing awareness of the importance of social and economic development, in some sense of the word. Lord Lugard[5], one of the earliest explorers in Africa and a visionary of peaceful colonial settlement, developed the idea that Britain was called to discharge a double trust; to develop African lands for the benefit of the inhabitants, and also for the benefit of the world at large, which needed their products – the so-called 'dual mandate'. The time was ripe for a forward policy of colonialism but it was not to be a policy of exploitation, he argued. The inhabitants of the territories were to share in the achievement and in the resulting benefits. Education, health, and other social services had, therefore, to be provided, both for their own sake and to enable the people to play a part in the new era. He believed that the resources of modern science had to be mobilised in order to intensify the war on the diseases of men, animals and plants which were hindering progress, to improve the quality and quantity of colonial products, and to promote their use in manufacturing and industry. The government's trusteeship for the African population was enshrined in the Devonshire Declaration in 1923 which pronounced that

'Primarily Kenya is an African territory. The interests of the African native must be paramount'. A new Native Affairs Department was set up to this end under John Ainsworth, a bulldog-like man and determined believer in the colonial role of trusteeship and the central importance of developing the indigenous population. In parallel, the importance of agriculture as a means both of raising the standard of living of the indigenous people, and of producing materials and goods for the home country was becoming evident.

Thus, there was a growing recognition that while a background in classics provided men with a powerful intellectual base, it might need to be supplemented with more specialised and technical knowledge, and in 1926 a full year's postgraduate training in Colonial Administration – specifically Tropical African Administration – was introduced at both Oxford and Cambridge. Whereas in pre-war times, technical departments (educational, agricultural, forestry and veterinary services) had to be content with a lower personnel standard than the administrative service, 'it is now recognised', mused a Colonial Office official, charged with recruitment for the service, 'that many of these departments need men of similar standing in this respect to the administrative staff, with the addition of professional qualifications... and the problems call more and more for the application of a higher degree of scientific knowledge and skill.' He goes on:

> The modern generation has no links with that dim past when a common conception of 'the Colonies' was of remote red patches on the map, the hunting ground of the adventurer or the purgatory of the prodigal son. The old catchwords 'white man's grave' and 'white man's burden' have lost their currency and meaning, and today the Colonial Service is esteemed as an honourable career of public service second to none in importance and prestige.[6]

Not only was technical support now seen as essential, but the type of men sought for such posts needed many skills and a strong temperament. Said Major Furse, Private Secretary for Appointments to the Colonial Service:

> The type of man required to make a good agricultural officer is of a very high order. Not merely must he be a highly qualified technical expert fitted by his scientific training to grapple with and overcome the problems that will arise in the work of his Department... but he must be endowed by his general education and upbringing with the tact and administrative ability necessary for dealing with the native mind.[7]

In 1919, a committee chaired by Sir Hubert Read concluded:

> Agricultural conditions in the colonies call for skilled scientific assistance on a scale much larger than heretofore.... The most important thing to be kept clearly in mind in this connection is the paramount necessity of securing the right type of man... the service of any but first rate men is not only uneconomical but may be expected to lead to disaster.[8]

By 1920, all the larger colonial territories had a government department focused on agriculture.

So, with this changing perspective, the cadre of colonial agricultural officers developed quite rapidly in the early part of the twentieth century. By 1925 there were about 150 agricultural officers in the whole Colonial Service, supported by a further 150 specialists and technical teams, but still relatively few compared with the administrators. Kenya was typical: of a total colonial establishment in Kenya of about 900 in 1928, the Agricultural Department counted 29 of whom only seven were agricultural officers or assistant agricultural officers. There were also veterinary, forestry and plant breeding services, a further 100 or so officers. The Department of Agriculture was seeking to improve the quality of its staff and in 1928 the head of the department was lamenting 'the paucity of suitable candidates' commenting that 'only young and inexperienced young men can be recruited for new posts and some time must elapse before they have acquired experience and knowledge of local conditions'. But by now there were new training programmes for young recruits to the Colonial Service, and the courses in tropical agriculture at Wye College in Kent[9] and Seale Hayne Agricultural College in Devon, both leading institutions, offered strong technical education to aspiring colonial officers. A specialist one-year course in tropical agriculture was established at the Imperial College of Tropical Agriculture in Trinidad. The process became more formalised and by 1934 a Colonial Agricultural Service was established, with formal academic requirements – at least two years of a relevant science subject plus specific training in agricultural practice and policy in the tropics.

So it was that the climate began to favour the combination of intellectual strength with technical competence and the door was open for men like Owen to play a role in the forward movement of colonialism in the new era.

And it was not an unpromising career path. In most respects, the terms and conditions for young officers were good. Salaries were reasonable but not particularly high by modern standards; Owen's starting salary in 1928 was £372 per year, equivalent to about £19,500 in 2010 prices (and after a 20-year career in Kenya it had risen to the present day equivalent of £42,000, which is not particularly high for a Deputy Head of Department, his final post).[10] But housing and a staff of cook and/or houseboy was provided as was home leave of six months every three years. Office hours were – officially at least – not arduous and were typically six hours per day (often 8.15am to 12.30pm and 2.15pm to 4pm) from Monday to Friday and often Saturday mornings. When on safari hours might be longer, and officers received an allowance to cover costs on the road – including the cost of accompanying staff – but no compensation for additional hours worked. You could retire at 40 with a full pension and expect to return to England with valuable skills and experience to offer for a second career.

Owen's background was typical for the advisory, technical roles. His family was devoutly non-conformist, and, while his mother was descended from farming stock in Cornwall, both parents were teachers. Alice Tyack, his mother, had been educated in the village school in Sithney in western Cornwall, and following the practice in the late 1870s became a pupil teacher at about 14. She moved to London where she met William Joseph Sunman, and they married in 1895. He was also a teacher and they settled together in Bromley, Kent, although she was always fiercely Cornish and named their house Pedna Vounder, the name of the farm where she was born and had grown up. She always mentioned her very Cornish maiden name given the chance (at the age of 91 she had a fall and when asked her name on admission to hospital she replied 'Sunman née Tyack'). Summer holidays were spent in a cottage in west Cornwall, near her roots. Their family was small; although Alice had been one of 13 (11 survived to adulthood) and William one of six, they had only two children: Jossie born in 1898 and Owen, born six years later.

Owen went to the local primary school, Wharton Road, and gained a scholarship to St Olave's Grammar School in Southwark. This was a prestigious school with a strong academic record. A list of the successful alumni shows a number of senior civil servants, judges and clerics; one eminent contemporary and close friend of Owen's was Leonard Beecher who himself went to Kenya with the Church Missionary Society in the late 1920s and eventually became Archbishop of East Africa. From St Olave's Owen won a major scholarship in 1923 to study agriculture at Wye College, London University, gained a BSc (Agric London) in 1926 and followed this up with a Diploma in Agriculture.

He was something of an individualist, possibly even contrary, and while in later years he lamented that he was a jack of all trades and master of none, a good all-rounder, his school reports suggest he had been expected to achieve a scholarship at Cambridge. Indeed, the reports indicate that the school was disappointed that he did not attain the necessary levels of success at school, so it is possible that the Colonial Agricultural Service was a second best option. In fact, it seems that his academic achievements were less than had been expected of him, as in early years he had shone in mathematics and passed his General Schools Exams with distinction. But he was active in many less academic areas; he was sporty, winning the one mile race in five minutes twenty seconds, and the school magazines show him as secretary to the school orchestra and an exceptional musician. In the school cadet corps he rapidly achieved promotion to corporal and was an active member of the Natural History Society. Despite his less than top notch academic achievements, these activities would have fitted him with the qualities Furse was seeking:

> you will have noted (he writes in his advice for interviewers) the extent to which your candidate was a leader or organiser at school or university in games, clubs, societies and so on; to be... a cricket Blue is in the long run a fleeting distinction, yet it may be at once a test and a presage of character.[11]

His failure even to sit for a scholarship to Cambridge may well have been a disappointment to his teacher parents, but his decision to go to study agriculture seems to have been firmly made. One can speculate as to why. Was he already thinking of the colonies? His older sister was becoming a teacher; perhaps there was an element of rebellion and a wish to spread his wings which led him in this direction? He had shown interest in natural history and science at school, but the study of agriculture at an out-of-London branch of the University of London was not an obvious second best – or next best thing – to a place at Cambridge. To go to Wye to study agriculture was a positive decision, perhaps influenced by the Cornish family holidays and the Cornish farms of his grandparents, perhaps motivated by the opportunities for travel and service that the subject offered.

Certainly, in choosing to go to Wye he had selected one of the two elite agricultural colleges, and the one with the stronger curriculum in tropical agriculture. The other was Seale Hayne Agricultural College in the West Country. Both were set up originally to look at farming in the UK but, as the colonial responsibility for agriculture developed in the colonies, Wye became a specialist centre which was the forming ground and home of kindred spirits for Owen. (It was not a fashionable choice and he received a fair amount of teasing; his former school mates referred to him as 'down among the farmers' in some catch up 'where they are now' type notes in the school magazine.)

Wye College was very pretty, a Grade 1 listed building from 1894, built in traditional brick over flint with Gothic windows and arches; wisteria, ivy and a herbaceous border lining one wall of the principal quad; romantic and charming with mullioned windows and linen fold panelling. Linked quadrangles give a medieval, ascetic air to the college and its intellectual and social life was rich. It had a long pedigree, founded in 1447 as a Latin school and seminary by John Cardinal Kempe, Archbishop of Canterbury and Lord Chancellor, and had been a centre for learning ever since. The village of Wye sits in the beautiful rolling North Downs of Kent, with a large crown carved out of the chalk by students for the coronation of Edward VII in 1902, high above the village. There was an experimental farm attached to the college, lying amid the valleys of hops and fruit trees, a classic rich Kentish agriculture.

So, in 1923 he arrived at Wye College. He loved it for its style and intimacy, and he met many like-minded souls there. He went on a college trip to the wheat plains of Canada and a picture shows him at the centre of a grinning boyish group of young men about to embark on that particular adventure. He plunged into college life with an enthusiasm he remembered with affection for the rest of his life. He set up a college orchestra – in 1923 his contemporary Stuart Gillett had lamented, as Secretary of the Music and Drama Association, that the college had no orchestra; by the following year Owen had established both an orchestra and a choral society. In athletics he bettered his previous one mile time to five minutes, seven and three-fifths seconds; took part in Hare and Hounds; and played rugby – he comes across as sociable, organising and energetic. He was on the committee of the Debating Society, and some of the motions provide an insight into the thinking and issues of the age. He chaired one debate where the motion was 'That Great Britain deplores the depletion of the cream of Great Britain's manhood through colonisation'. The arguments ran to and fro – one participant commenting that there was an inherent national tendency to roam the world – and that it had now become so easy that it drained the country of good men. The opposing arguments were mainly economic – a brain drain which undermined the home economy. The proposers noted that after the Napoleonic wars it was thanks to colonialism that Britain recovered. The motion was defeated 21/36. Another motion proposed that 'Nationalisation of land, mines and railways will be beneficial to this country'. It was defeated unanimously.

Each term had its own social events: Christmas dance, spring dance; one or two 'Smokers' each term; plays and concerts. But the highlight of the year was Cricket Week, held in the third week in July. Alumni returned, the students organised the sporting events – not just cricket but athletics, swimming, tennis as well. On the Friday night the Cricket Week Dance was held – a splendid affair. A marquee was set up in the main quadrangle; the gardens were manicured to perfection; roses draped around the doors. The dance began at 10pm, with a London band playing until 2am – then the college bands took over until 3am or later. Breakfast in Folkestone and sea shore walks, a motor ride in the summer dawn – it was romantic and beautiful.

The syllabus was practical rather than theoretical – papers published contained such titles as 'Guide to the farms, livestock and poultry, orchard and vegetable ground, hog-garden and forestry' (dating from 1913); 'The economics of fruit farming' (1919); 'Report of the cost of food, in the production of milk, in the counties of Kent and Surrey' (1908–10). But by the mid-1920s it had become a world-class centre for agricultural research and by the time Owen arrived it was the principal training ground for the growing agricultural service of the colonies.

During Owen's time there, an Old Wye Man, Alex Holm, was Director of Agriculture in Kenya, and he came on a recruitment trip to Wye in the summer of 1924. As the debating topics suggest, men had gone from Wye to the colonies in significant numbers, but most had gone to work as farm managers or coffee growers among the white settlers. It was not until 1923 that the department began to turn its interest to the development of African agriculture, under the leadership of Alex Holm, and he came to seek suitable young men to work with him in this field. The three he persuaded to join him would stay for a long time. Owen was one. Another was Ted Gaddum, who like Owen was musical and sporty – he took part in music and drama and was a founder member of Owen's orchestra. He went from Wye to study for an agricultural diploma at Oxford and then on to take the specialist one year course in tropical agriculture which was established at the Imperial College of Tropical Agriculture in Trinidad, and thence to Kenya. The third was Stuart Gillett, who would ultimately become Director of Agriculture in Kenya. He was clearly an organising and organised young man. He sat on several committees: the Committee for Music and Drama as noted above, but also the Finance Committee, the Cricket Week Committee and the Central Committee; he was captain of hockey and, once he had arrived in Kenya, he rapidly became engaged to Alex Holm's younger daughter whom he married in 1931.

They were not the only Wye men to go to Kenya from Owen's cohort: Horace Taylor, arrived in 1930, and other contemporary names who would crop up over the coming 20 years were Alan 'Granny' Ransford, C. G. Bishop, Bill Leckie, Croft Handley and Roger Lambert. (The Wye men liked nicknames; Owen was known as Sunmaid – presumably to his annoyance.) In a letter written from Kenya in 1931 Ransford notes that there were eight Old Wye Men in the department, and a further nine or ten working elsewhere in Kenya, on private estates. There was a strong bond between the Wye alumni. There is always a special feeling about being in a group of pioneers, sharing new experiences and privileged position in the history of the college and perhaps the world. Another bond perhaps lay in the fact that they were the elite students who were studying for degrees, surrounded by local farmers aiming for agricultural diplomas. Whatever the reason Owen felt a lasting rapport with his contemporaries and, once he arrived in Kenya, he thrived, as he had at college.

He was met off the train in Nairobi by Alex Holm and was accommodated in government apartments – a photograph shows him relaxed in shorts sitting on the balustrade of the apartment block, already at ease in these new surroundings. New cadets and technical officers normally arrived in Kenya with no idea of when or where they would be posted; some were whisked straight off to remote stations; others stayed in Nairobi for some orientation first. This was Owen's case; he spent some weeks learning the business of the Agricultural Department in Nairobi, at Head Office in Nairobi and at the Scott Agricultural Laboratory in Kabete, an experimental station and one of two training schools for Africans, learning the specialist's role. The laboratory was one of two, established as training colleges for Africans in the early 1920s, offering two year courses for extension workers and agricultural instructors, to work with European agricultural officers. The second training school had been set up at Bukura in Nyanza, on a site which included provision for nursery sites and the development of cotton, and is now the site of a very large, still predominantly agricultural, University College. The Scott Laboratories, now called KARI (the Kenya Agricultural Research Institute), also continues the earlier work and is now a centre for the analysis of impacts of climate change.

The policy focus for African agriculture was to improve food crops through improved husbandry and, for example, application of manure, and to develop cash crops for export. The African agricultural instructors were central to this strategy of improving agriculture in the Native Reserves through example and teaching. By 1928, over 40 had been trained, working as instructors, mission teachers (gardening was taught at several mission schools) and plant breeding assistants and the courses were extended to three years. At the schools pupils undertook elementary school work, field work and cultivation of gardens, and African agricultural advisers were trained to manage demonstration plots, as traditional crops – sorghums, millets, peas and beans – were beginning to be replaced by faster growing varieties, and cash crops such as cotton, wattle, sesame seed (*sim-sim*) were introduced.

By 1930, the Department of Agriculture maintained 106 demonstration plots throughout the Reserves, and seeds bred in Scott and Bukura were purchased by the Agricultural Department for distribution to African farmers. Local Native Councils (LNCs) also voted funds for the purchase of seed for their areas, and according to annual agricultural reports there was great interest in the purchase of Rhode Island Red cockerels, to breed with local poultry. LNCs had been established in 1925, with budgets and revenue raising powers and were the principal instrument of local government. Non-elected, they were formed from local chiefs – cynics would say their function was to do what the DC – the president of the council – told them. But they were more constructive than such an interpretation suggests. They were seen as business-like and according to the contemporary historian Margaret Dilley, writing in 1938, 'their work has come to be the most important factor in the improvement of Reserve conditions'.[12] They levied taxes on the African population, spending the revenues on education facilities and schools, agriculture, roads, medical supplies and water supply – in other words the normal range of local government concerns. Dilley notes that 'their willingness to tax themselves to provide services has proved embarrassing for Government, which has been unwilling to allow natives to fund permanent works without some Government contribution to maintain central control'.

Extension work in the Reserves included selective breeding of other livestock, introduction of ploughs and other iron implements and the introduction and management of irrigation channels. The annual Agricultural Department reports provide great detail on the types and quantities of crops – quick maturing yellow and white maize for example – which were being introduced. Agricultural officers worked closely with the administration staff:

> [We should] mention the very great interest taken in native agriculture by many Administrative Officers, and would recognise the great help and encouragement they are [sic] to Departmental officers in the field and to the Department

> itself. There is now a generally unanimous demand for the services of Agricultural Officers by Administrative Officers...[13]

During the brief orientation phase in Nairobi the newcomer would have learned the background, how the Department for Native Agriculture had been set up in 1920. He would have been told how 'the Department is becoming more and more interested in the development of the highly fertile Native areas', and how it was the department's aim that the natives should take 'a proper place in the economic life of Kenya, they should be *made to utilise profitably* (my italics) those large sections of the country in which they are left undisturbed in peaceful possession'. He would have learned about the training of local inspectors, about different crops, different problems with crops, soils, markets and cattle breeding.

And orientation included the energetic social life of Nairobi – theatre, music, the elegant Norfolk Hotel and its verandahs, and the Nairobi Club, haunt of the government staff. In practice, most of Owen's contemporaries were posted in the provinces in Kenya[14] and in July, after about five months in the country, he took over the post of assistant agricultural officer in Meru, a large Native Reserve to the south and east of Mount Kenya, in Central Province.

The following extract from a departmental report could well have been the briefing Owen received before taking up his post.

> Meru Reserve, eighty miles long with its 160,000 inhabitants, its variety of climate and elevation from 2,000 feet to 7,000 feet, its plentiful rainfall and numerous streams and its very fertile soil, has a vast potential agricultural wealth. It must be remembered that from an agricultural point of view, development has scarcely yet begun. The country is ripe for development, communications and transport are becoming easier and many of the native [sic] chiefs are showing themselves willing to cooperate with the Administrative Officers in an agricultural policy which will lead to increased productivity and prosperity.

> Such a policy must be far seeing and consistent from year to year, and needs, for its pursuit, an increase in the staff of native instructors and a judicious but generous expenditure of money.[15]

Meru itself is an area of great beauty, dominated by Mount Kenya which rises to over 17,000 feet, but it was remote, 200 miles from Nairobi, and 75 miles from the railhead at Nyeri, round the north of Mount Kenya. The journey would have taken a few days: a train to Thika and then on to Nyeri, the capital of Kikuyu, a journey of about 80 miles; overnight in Nyeri, or the Silverbeck Hotel in Nanyuki, some 35 miles further (and quite close to the famous Treetops Hotel). The road from Nanyuki to Meru was slow, especially during the rains. But it was a wonderful and exotic first posting, appealing to the romantic in Owen's nature.

The wildlife and forests of Meru were exceptional, and there is now a national park, based around the Giitune Forest, which contains many indigenous tall trees and bird species. Now recognised as a World Heritage Site, the forest then was thick and dark, much associated with the myths and belief systems of the Meru. The forest was home to the tree hyrax, which looks like a guinea pig but has an extremely raucous screech, to monkeys, elephant, crocodiles and lion. Deep in the forest are twin waterfalls, remote and hidden, scarcely visited now, on the river Kathika, and rarely seen by white men – at least before 1930.

An exception was the first DC of Meru, a man called Edward Horne, known locally as '*kangani*' or wanderer, because he travelled widely in the district on the back of his white horse and penetrated deep into the lands of Meru. He was a man of great energy. Short, stout and hot tempered he was only 27 when he came to Meru, after a career which had already involved a conventional British education, several years as a lumberjack in Canada, and participation in the British punitive expedition in Embu sometime earlier, where 'spitting sticks' were used to tame the recalcitrant tribe. The Embu experience encouraged a more receptive response from the neighbouring Meru. Horne's arrival in Meru was more peaceable and he dealt with the traditional Meru chiefs, giving one, Mwende Mbogorio, a black blanket, the symbol of the office of chief and authority of the colonial government, in exchange for a cask full of honey and a ram, the traditional symbol of peace in the Mount Kenya region. With Mbogorio left in charge, Horne consolidated his position and set up camp, a development which was welcomed by the Meru since the camp (reinforced by spitting sticks), provided a barrier to cattle raids from the Masai and Ndorobos – the traditional hunter gatherer tribe living in the forests of Mount Kenya. The story goes that he planted a flag in an area of open meadow, on the site of what is now the Pig and Whistle Hotel, and declared that the area within a radius of one mile was Meru township.[16]

Horne was a fanatical big game hunter, but also had a talent for languages and during his time in the Colonial Service mastered the Swahili, Gikuyu, Meru, Embu, Masai and Nandi languages. He was also an exceptional engineer and, with Rodney Stone, DC at Embu during the First World War, designed and built the main road from Embu to Meru; it was at the time extremely well graded, and it is said that Stone and Horne used no instruments at all – and it is the basis of the 'round the mountain road' which was useable in Owen's day. It was a wonderful road, swooping and plunging through dense forest and river valleys and so well engineered that it is the foundation of the present road.

The *boma* – as the government areas were called – was considered one of the best by the young recruits. It had a good climate and was generally healthy – and it was well laid out on a grassy downland (looking, said Owen, like an English park) with clear views across to the Jomboni Hills and the ever present Mount Kenya. Behind the *boma* was a thick belt of forest, with herds of elephant and the screeching hyrax. (The term *boma* was originally an enclosure made from bent over thorn bushes and branches, an enclosure for cattle, but it came to mean the fenced off area surrounding government housing and offices, and eventually came to refer to the government area in any town, fenced off or otherwise.) The Pig and Whistle Hotel lay down the hill from the government *boma*.

In those days housing was fairly primitive: wooden constructions; external long drop latrine; oil lamps and often earth floors, which needed to be constantly swept and smeared with dung to keep jiggers – little insects which penetrated the soft skin around the toes and caused terrible irritation – at bay. Most of the old *bomas* had been erected by district commissioners in the early days with precious few funds. While they could be considered a marvel of construction under the circumstances, by about 1928 they had passed their best, and there was a further problem that even before the arrival of the agricultural officers, there were only just enough to go round. Owen was only the second agricultural officer to be appointed to Meru, and his house was very dilapidated, of an old hexagonal 'romantic' (surely a euphemism) timber construction which had been built when the *boma* was first established 20 years previously. A nearby earthquake in 1928 caused considerable damage to the houses – and one of the walls had subsided by about a foot, letting in daylight around the top. Some of the young recruits were put up in stores or offices which had been converted to houses once they were deemed unsuitable for their more formal purposes (that at least is what some wrote home). In fact, a new house was built for the agricultural officer during Owen's stay at Meru, a wood and brick structure with a roof partly of corrugated iron and partly of a thatch of banana leaves and stems, still an earth closet but hot water provided by a wood or charcoal fire under a large former kerosene can. The house became known as the 'Old Flying Horse' the name of the pub opposite the College in Wye, as four successive Old Wye Men were posted there – first of all Croft Handley, who preceded Owen; after him, Ted Gaddum, and 'Granny' Ransford. They felt like pioneers with a great shared camaraderie.

Even if the houses were not of very high standard, all *bomas* had some sports facilities attached, often a golf course and tennis courts. Within the terms of employment, it was the general belief that after a day's work in the tropics, it was important to have relaxation and exercise, so before sundown many officials developed the practice of a game of tennis or a round of golf. *Mens sana in corpore sano* – a healthy mind in a healthy body. To this day, the geographical proximity of the local club – with its sports facilities – to the government homes and offices is evident.

Remote as it was, there was a community of Europeans in Meru by the time of Owen's arrival. The government *boma* consisted of half a dozen officers, and there was a small detachment of the King's African Rifles (KAR) nearby. A Methodist mission with the standard school and workshop was some ten miles away, there were a couple of Italian missions, and there was a saw mill operated by two Europeans, Aubrey Young and Pat Ayre. Ayre was a white hunter, and qualified to conduct large shooting safaris – he accompanied the Duke and Duchess of York (George VI) on their honeymoon safari to North West Frontier in 1924. There was plenty of game shooting to be had, too. Rodney Stone was still DC and, a keen hunter himself, took Owen with him on shooting expeditions. Owen writes of hunting elephant, rhino, and buffalo, 'although' he notes 'as far as I know I have not shot anything. But bird shooting provides good sport and is great help in the larder, especially when camping'. (Later he records shooting his first buffalo.) Hunting was only partly for sport, since elephant and buffalo were both very destructive and were hunted to protect crops.

So, although it was 200 miles from Nairobi and 75 miles from the railhead, there was enough social life. There was a nine-hole golf course and a hard tennis court. In the rolling country, with plenty of rainfall and sparkling streams the trout fishing was very good and there were enough Europeans around to ensure regular bridge parties and supper evenings. It was a popular posting, and Ted Gaddum liked it so much when he arrived a year later that he promptly sent for his wife to join him. The 'round the mountain road' ensured visitors from Nairobi too. The Pig and Whistle was very popular at the time, and during the 1930s, it seems to have been a glamorous location, much used for movie making. People were warned that it was prudent to telegraph ahead from Nairobi if you wanted to visit for the weekend since it might well be full of film crew. Owen commented on the golf course ('very good') in a letter to his alma mater, Wye, but added that 'there is no piano'. Still, he wrote 'people here a good crowd, and on the whole have been here longer than usual. Constant changes tend to make things unsettled and have a bad effect on the policy in any one district. Efforts have been made to keep transfers down to a minimum'.

The DC's office, where Owen and his colleagues worked, was the first stone building in Meru. It still stands, a typical colonial single storey structure with polished wooden floor and corrugated iron roof, with two wings flanking four wide steps to the wooden verandah. It now houses the Museum of Meru with exhibits of traditional Meru village life and animals from the forest. Across the road from the DC's office is the old Law Court and behind that the prison, which was built of brick in 1926 – and is still used as the town jail. The old Law Court now houses a Kenyan NGO, the Rural Community Development Agency, which works with girl victims of HIV and other abuses. The Pig and Whistle is still a hotel, a statue of a bewildered looking pig with a broken whistle in its mouth, at the centre of the gardens.

Day to day work for the agricultural officers was a round of planting, lecturing, exhorting. Demonstration plots, better seeds, new techniques – all formed part of the armoury of the agricultural officer, and they were supported by their African assistants who were posted in various communities in the district. In his first annual report Owen describes all the crops being grown and the progress with them. He hoped that a constant supply of good seed would be available to build up a surplus of maize on the Reserve. Similarly, beans were grown throughout the Reserve, and were an important staple crop – but the yields were poor and the beans low yielding, so that 'the greatly superior Canadian Wonder are being introduced'. They worked to introduce maize, beans, wheat, rice and English potatoes. He notes that maize grew well in the area, but little was actually grown and the traditional strains tended to be very small. He writes with enthusiasm of his experimental and seed bulking plots:

> I have opened up a new plot at the *boma* of some four acres with room to extend next season. There I have planted plots for demonstration of maize and beans and wheat. These are the crops which we are trying to get the native here to grow as staple food crops, being so much heavier yielders than their native crops. In addition, I have laid down various trials of rice, varieties that should be able to grow on our higher slopes, and cultivation trials of maize and beans and trial plots of peas and flax. On other plots in the district we are growing sim-sim, potatoes and buck wheat, according to altitude, which varies from 1200 ft to 8000 ft.[17]

Owen went on safari on motorbike with his cook/houseboy to visit the agricultural instructors and local chiefs in the various corners of the district, to demonstrate techniques and discuss markets with local farmers. At an official level, the government wanted to develop cash crops for the local farmers, to improve their standard of living; but Meru was so remote that there was little profit in cash crops. In 1927 the main concern for the agricultural officers was to find crops which would stand the journey to the railway. Some cereal prices were reasonable at the time but native maize was worth about 7/- a bag and *wimbi*, the local staple fetched only about 4/- which would barely have covered the costs of transport to the railhead.[18] So in practice, Owen and his team of five African agricultural inspectors worked to encourage production of better food crops, in the understanding that a bountiful supply of food is essential to the peaceful development of the people, and that once assured, other and more valuable crops may in turn be produced in increasing quantities, as he wrote in an annual report in 1929. Practical examples included the application of animal manure to vegetables but also the introduction of metal tools and ploughs. Ted Gaddum describes training six oxen to act as a demonstration plough team.

The safari was at the heart of the agricultural officer's job. In his report for this first year Owen notes that 'I have travelled 1,513 miles by motor-cycle since July, and have been out of station for 71 nights in the Reserve and 6 nights in the Northern Frontier Province on Government business'. The following year he was away from the district for 22 days of leave and eight days on duty and away from the station on duty in the district for 98 days; and he travelled 2,272 miles by motorcycle in the district. Travelling by motorcycle and camping must have been a challenge, with an agricultural instructor on the pillion, but tremendous fun. 'The banks of the rivers were clad in maiden hair ferns, orchids hung from the boughs above, and always, on turning a corner, there was a different aspect of Mount Kenya' writes one local European. The bush was full of rhino or other wild animals, so the traveller had to be constantly aware of his surroundings. The terrain was hilly and it must have been exhilarating and exciting – the open spaces, the forests and rivers, safaris to remote demonstration farms, cultural challenges and the feeling that you were doing something worthwhile.

Life was not predictable. A colleague, the DC while Owen was there, tells of safaris in the mid-1920s in the Nyambeni Hills, which run through Meru District. On one occasion, while he was camping at the Tana River, a local begged the DC to accompany him to the Tana falls, some miles away and shoot a crocodile that was haunting the place. He agreed, but was told he must first walk through a sacred grove, taking care not to carry any sharp instrument, for fear of damaging the trees. They arrived at a small deep pool some distance from the river, where there was no sign of the crocodile; then the man pushed a long stick into the depths – until the croc was finally disturbed. A few bubbles appeared, then a malevolent eye, until it leapt out of the pool and scuttled away towards the river. The DC despatched it with a single shot.

But it was also an uphill struggle in terms of the ambitions and hopes officers had for their area. The reports write of the setting up of agricultural experimental plots, and Owen's first reports are proud of achievements as well as hopeful that more will be developed. But the local population was often very suspicious. It was widely believed by the natives – particularly the Kikuyu, but also the Meru – that the aim of experimental plots was to find out where the best land was, so that the government would then expropriate it for European farmers after the local Africans had broken the soils. So the local chiefs would offer the worst plots possible to the Agricultural Department, so that the results would be poor, the chiefs having safeguarded the good land for themselves. The young agricultural officer on safari needed to build a trust with African farmers.

The Old Wye Men had visions of teaching willing learners, enough resources to provide good lessons, helping people to understand new ways of doing things, new ways of marketing produce. But calamities like locusts, famine and drought – a series of terrible scourges across Kenya in 1930 – meant that the officers' main efforts were in trying to deliver enough emergency foods to keep the population in remote areas like Meru alive. Carefully planned rotations and demonstrations fell by the wayside as the officers and their assistants struggled to get food supplies to the impoverished population. The locusts arrived in January of Owen's first year in Meru, wiping out about three-quarters of the crops which were then nearly ready for harvesting.

An invasion of locusts could sweep through the land at astonishingly short notice and in vast quantities. To quote Mervyn Hill:

> In the afternoon, all may be well, the crops ripening to harvest and the cattle grazing contentedly. Then the sky darkens, as a vast swarm of locusts, miles across and miles deep, threatens from the horizon. At first it looks like a dirty smudge, like the drifting smoke of a forest fire or the gloomy murk of a sandstorm. Soon the locusts fill the air and man is virtually powerless to prevent their landfall if they so wish. And if a large swarm settles for the night, there is little worth harvesting on the morrow and little for cattle to eat. Apart from their theft of the cattle's grass, they lay the earth bare to excessive erosion by wind and rain, and in their poisonous path grazing is slow to recover.[19]

Trains were literally stopped in their tracks as locusts made the rails so slippery that there was no traction at all; cars were forced off the road. Photographs show how they swirled around, getting in your ears and eyes and mouth, intent on stripping any plant life that came their way.

It is almost impossible to imagine the intensity and horror of a locust invasion. They were hugely voracious; a locust needs to eat its own weight in food each day. The most menacing locust is the desert locust, *Schistocerca gregaria*, gregarious as its name implies and travelling in great swarms. Owen describes how a swarm arrived where he was staying under canvas during the night, with the leaders passing on and others sleeping in the bush nearby. He writes 'This morning, since 7am, there has been a steady stream of them, flying along this valley at about 12 or 15 miles per hour; it is now noon, and there is no sign of the end. This makes the swarm 60 miles long!' They would consume literally tons of crops and grazing each day, and when they had passed over, the land was left desolate, stripped of crops and grazing; a stubbled, leafless vista where hours before food enough for the coming year was growing in peaceful *shambas* and fields.[20]

To have any hope of controlling this biblical plague, the trick was to try to catch them and burn them before they hatched into flying insects, while they were still hoppers. If they had already hatched into flying locusts, the villagers would shout and clatter, trying to make a lot of noise to discourage the locusts from settling. In Embu and Meru there was a huge operation to destroy hoppers with 28 trucks and 48 officers in the field at the height of the plague. But the locusts returned and returned again.

Inevitably, inexorably, the locust invasion was followed by famine. The role of the agricultural officers was not now to encourage better farming by Africans content with their longstanding practices but was to supply food to prevent the population from starving. Not only did the locusts strip the land of everything, but they were followed by drought. The plague emphasised the fragility of the lives of the African population. Once the rains started again, the roads were totally destroyed and efforts to bring in sacks of maize for the people were again thwarted. Owen writes in one letter back to the college that the collective result of this was that the famine was worse than ever; he was camping in one of the worst parts of Meru District selling the odd scraps of food they could get hold of. By the end of February the officers were importing food to sell in the Reserve. Prices soared, and food was sold to the African population on credit since cash resources had long since been spent on the inflated prices of the markets and *dukas*.

This was a long way from the prairie lands of Canada or even the flat brown beet fields of East Anglia, however good the training and education offered at Wye. Demonstration plots – rice and Canadian beans; better chickens; collecting better seeds for planting for the next year; this is what they all wanted to do – developing some 'real' agriculture, hoping for things to go better in the long term, in the future. The height and rainfall of the slopes of Mount Kenya should have provided good potential. But the reality was that lives were hard and livelihoods so fragile that well-planned programmes were often swamped by the immediacy of human need.

There were some attempts to develop more commercial farming for the Africans. During 1929 Owen was optimistic about a scheme to provide for the development of commercial *shambas*. The idea was that large tracts of land would be broken up by mechanical means by Europeans, and then cultivated and planted by Africans, with both the Africans and Europeans sharing the profits. Thus, Young and Ayre from the nearby saw mill agreed with a number of chiefs and headmen to prepare some areas of grassland for planting, while the Africans sowed and tended the crops; 50 per cent of the value of the yield would be paid to the Europeans for the ploughing operations. But the scheme was of course viewed with suspicion in some areas, where some locals feared that they were being exploited. 'This suspicion the Administrative Officers have been at great pains to remove, and once a satisfactory harvest is obtained there should be no difficulty in continuing and developing slowly along the same lines', writes the optimistic Owen.[21] It is not reported whether the scheme worked out, but this type of mutual incomprehension between development work as seen by European professionals and suspicion by the Africans runs through this colonial period; and underlies much of the tension which built up later to Mau Mau.

After 18 months or so, Owen moved to Embu, a small but long established *boma*, although he continued to look after Meru as well. There were only three officials there at the time, all bachelors, himself, the DC and an ADC. As he writes in a letter to college, this posed problems as they did not have that 'delightful combination of a "four", which is so essential in small communities for people who are keen on tennis, golf and bridge'. 'We do our best', he goes on, 'by incorporating the occasional visitor'. They also spent a fair amount of time reviving the old nine-hole golf course. Clearly there had been a succession of people in Embu who did not care for the game, so that 'hole after hole has disappeared'. The *boma* had a beautiful garden, set out with trees and avenues by an earlier DC and his wife. But it posed problems for would-be golfers trying to find space for a hole of 400 yards in length.

It is hard now to imagine the reality of life in remote central Kenya – days spent under canvas, sometimes with a British colleague, sometimes with African colleagues or a servant; the suspicion of the Africans; the prevalence of disease; isolation and slow, irregular communication. Did the reality match up to the idealism and vision engendered in these young men during their days at Wye? Leaving Wye, full of promise and expectation for Kenya, a country full of potential, of people wanting to be helped, to develop, to improve their standard of living. The reality was much tougher, and they found themselves instead working with a native population whose ignorance and conservative social structures bred suspicion and sometimes hostility. Masefield, the only historian of the agricultural service, comments that despite the glamour and excitement, the work of the agricultural officers was not easy. Their work was:

> handicapped by scientific isolation, trying climates in nearly all areas, frequent spells of ill health. The adoption of improvements which they were trying to stimulate among

> the farming population was held back by massive ignorance, archaic social organisation, and in – as emerges later – some territories as independence approached by political opposition. The subordinate clerical and field staff available were poorly educated. Money, except in the final years of colonial rule was always short.[22]

None of the letters – his or others' – suggest fear or anxiety or even much frustration, but then what young man would show these emotions on letters home to mothers or lovers? The density of the jungle, the weather, the intense rains and the cold – under canvas and accompanied only by an African agricultural instructor – it must at times have been enough for a man to question the wisdom of his career choice. But we cannot hear that in the cheery totalling of the number of miles travelled by motorbike in the annual report, and when he went home on his first home leave in 1930 what tales he would have told his adoring parents – and their neighbours: camping on open plains; hurtling his old motorbike along dirt roads or surveying the land on horseback; tennis; golf; wonderful scenery; new and unimagined experiences; and the family of colleagues from Wye.

The desire to develop and improve the standard of living and prosperity as a basic aim shines through very strongly. The role of the agricultural officer, and indeed most of the administration up-country at this time, was to develop the land as well as possible. But the Agricultural Department was taking on a prodigious challenge. Meru had a population of 160,000, the province was 80 miles long, and the staff of the department consisted of one newly arrived European and four agricultural inspectors. But they tried: teaching and encouraging, explaining and helping government officers and missions combined to become a driving force for development. The missions provided education, itself a great force for change and development, and the practical inputs and technical support – from the agricultural service but also in the veterinary fields and forestry – enabled changes in people's standard of living and the transformation to the modern world.

Many of the officers saw themselves as holding a trust: 'I see myself as a guardian for the local native until he is ready to take over for himself' wrote Oscar Watkins when he arrived as an administrator in about 1910. Oscar comments that the colonial officers represented a type of 'muscular Christianity' where the sense of doing good and improving life for the African was important. The sense of developing the country and building up its capacity was inherent in the work of the government, and particularly so in the Native Reserves. Looking back on the changes, some were highly dislocating and difficult. The switch to a cash economy from one based mainly on barter and subsistence caused great shocks and upheavals – but change was inevitable, and not always unwelcome.

But, even if they had anticipated the difficulties imposed by geography and (always) scarce resources, would they have understood the wider tensions between settler and African, settler and government? And the extent to which the past defined their potential impact in the future? They arrived at a time in Kenya when there was no longer a *tabula rasa* – there was already a history on the ground. Had they been warned about this? The settlement of Kenya and its history had already formed a place and polity which would affect the ways in which these agricultural officers would be able to work, and what they could achieve. And the tensions between the settlers – their goals and achievements, both professionally and politically – and the government with its concern for the betterment of the African population – would provide a difficult backdrop for the Department of Agriculture and its new recruits which would conspire to limit what could be achieved. Extracts from the departmental reports and from letters from some of the officers describe a professional commitment to improving the lot of agriculture in Kenya – particularly native agriculture – but it was not an easy task. The idealism of the colonial vision would be tempered by the events on the ground of the previous 20 years.

2

THE BEGINNINGS

The side winds of policy[1]

> 'Kenya was conquered not by force but by a railway.'
>
> Governor Eliot, 1903

> 'There were three types of European in Kenya: missionaries who brought Christianity; the government servants who brought development; and the settlers.'
>
> Lukas Alebe, archivist at the Anglican Church of Kenya, 2011

By the time that Owen arrived in Mombasa in 1928 Kenya had been finally established as a colony for just eight years, but had held a place within the British sphere of influence for about 30. The world that he was entering had been moulded by the collision of several powerful forces; these framed that world and would affect the lives of him and his colleagues.

During the latter part of the nineteenth century, the European powers were expanding their interests in Africa and in 1884–85 the Conference of Berlin delineated areas of influence between the European powers. The areas of influence were seen as just that; they were by no means the plan for future countries that they have in fact turned out to be, with many problematic outcomes. The idea that Africa, this vast, rich and seemingly empty continent, would emerge into coherent nation states must have been impossibly far from the imagination of the actors in the 'Scramble for Africa'. The population of Kenya was sparse and the economies unsophisticated. There were between 2 and 3 million Africans made up of three main language groups – the subsistence Bantu in Western Kenya and around Mount Kenya, the pastoralist Masai of the plains, and the Nilotic Luo who settled around Lake Victoria – not that it was called that yet. The appearance of the land was of agricultural islands in the midst of a sea of pastoralists who dominated the open plains and grasslands and the dry north of the country.

Yet if nation states would have been unimaginable, European interests were nonetheless taking shape. Germany had strong interests in East Africa by the end of the 1880s, but the British government had largely maintained a hands-off policy in the face of German activities. Indeed, Britain had been resisting the idea of taking colonial responsibility in the area, but, as German interests in colonisation in East Africa began to strengthen, she began to feel the need to become more formally involved. In practice, if not policy, there were already substantial British interests in East Africa.

> 'The East African interior had been revealed to the world mainly by British explorers. The vast majority of the several hundred missionaries there were British. The Queen's Indian subjects controlled Zanzibar's economy through loans to finance the caravan trade and through near-monopoly of the coast's retail outlets. Furthermore, Zanzibar itself had long been a sentry box on the southern sea lane to the east, and it was very much in the Foreign Office's interest that British influence along the coast somehow be upheld as paramount despite the German intrusion'.[2]

There was, no doubt, a complicated mixture of motive and objective in the Scramble for Africa, including the search for markets and raw materials, evangelism, abolition of slavery, military glory and 'simple vainglory'. Margery Perham likened the conflicts in the area to a lot of greedy, quarrelsome children in a school playground; none quite big enough to dominate all the others, kicking and then making up to each other, sulking, coaxing, telling each other secrets and then 'splitting'. Britain, she suggested, was like the rather aloof child in the corner, unwilling to join wholeheartedly in the rough games and yet warily watching lest too many of the quarrelling schoolmates should suddenly combine against her.

Whatever the balance of motive, Britain was slow to step forward in East Africa, but in 1888 she finally took the step of administering her area of influence; not directly since government resources did not permit at the time, but through the indirect but tried and tested model of the chartered company. In 1888 then, the government supported William Mackinnon, an energetic and visionary entrepreneur in East Africa, in the establishment of the Imperial British East Africa Company (IBEA), which became the vehicle for British activity. One of the company's principal tasks was to win local recognition of the British sphere – they negotiated no fewer than 92 alliances with tribal rulers – while at the same time opening up the country to trade. Trading posts were opened in Machakos, east of present day Nairobi – the company's representative there was John Ainsworth who was to become the district officer for Nairobi and later the first commissioner for native affairs in Kenya – and at Mumias, a bustling but malaria-ridden caravan station by Lake Victoria.

But development was slow. The IBEA failed to show a profit and the idea of developing a railway to open up the interior began to take hold. Mackinnon made proposals to the Foreign Office that it should support the construction of a railway, and it was the railway which really stimulated the European settlement of Kenya.

Two principal factors underlay the investment in the railway. During the latter part of the nineteenth century, the movement to stop the slave trade in Africa had taken on great momentum. At a conference in Brussels in 1890 the principal powers engaged in the Scramble for Africa agreed to take various measures to put down the slave trade and control the liquor traffic. The best method of destroying the remnants of the slave trade was considered to be the linking of Central Africa with civilisation by means of railways and telegraph lines. At this conference, Britain promised to build a railway to the interior from the coast.[3]

Long-distance trade and caravan routes had linked the heartland of Africa with the eastern coast, Arabia and the Indies since time immemorial, and caravans moved goods and people – not only slaves – between trading centres. Caravans were a costly and dangerous means of transport, so that they would be used only for high value and reasonably portable commodities – mainly ivory, carried as head loads in great caravans – 2 or 300 men strong – by slaves on well-defined routes from the heart of Africa to the coastal ports. Slaves were also used in the great trading caravans, carrying goods through northern Kenya to Uganda, although in the main slaves were taken from the Congo, Rwanda and Burundi. There were several different routes, but all began or concluded through miles of harsh, inhospitable arid country. The ivory was of great value, but, in business terms, the beauty of the slave trade was that at the end of the journey the transport had value as well as the goods. At the coast, slaves were shipped on to Zanzibar whence ivory was traded on to the rest of the world, and slaves were bought by Arab traders and shipped to Arabia in Omani dhows.

The treatment of slaves during the journey was abhorrent. Livingstone estimated that, of every five slaves in a caravan, only one reached the coast. The passage of any slave caravan could be traced simply by following the vultures and hyenas that shredded the flesh from an endless winding line of putrefying cadavers; Livingstone, in his diary, wrote many observations such as 'we passed a woman tied by the neck to a tree and dead'... 'we came upon a man dead from starvation'... 'we passed a slave woman shot through the body'. Once sea borne to Zanzibar, conditions were no better. The slave dhows – less than 100 feet long – accommodated 600 slaves, crammed below decks in specially constructed bamboo tiers with about three feet of head room; where one neither sat nor knelt nor squatted, but achieved a crippling combination of all three. Meals on board were a daily handful of rice and a cup of stagnant water; and there were of course no sanitation facilities and disease was rampant.[4]

Zanzibar, with its famous slave market, was the centre of the trade. The conditions for the slaves improved at this stage, as they had to look fit enough to attract buyers, but it was too late for many. Britain passed the first Slave Trade Act in 1807, and slavery was officially abolished in the British Empire in 1834, but it took considerably longer to control the slave trade in East Africa than in West Africa. Treaties were signed with Sayyid the Sultan of Zanzibar and with the Omanis during the 1840s, but these had little impact on the volume of the trade. British warships were stationed off the African coast but they were few and the dhows were fast. By the mid-nineteenth century, opinion in the Americas, where the West African slave trade was focused, was veering towards abolition of slavery. In contrast, the countries receiving slaves from East Africa saw nothing wrong with the trade. In the Muslim Arab world and in the Indian subcontinent ownership of slaves was acceptable and normal, and slaves were used as household servants and for other aspects of domestic life, generally under reasonable conditions, rather than as indentured labour in plantation agriculture. So there was no pressure from recipient countries to cease in the eastern trade. In fact, the Act of 1834 exempted the East India Company, Ceylon and St Helena, and the eastern slave trade continued until the end of the nineteenth century. It was officially abolished in Zanzibar in 1897 when she became a British protectorate. Thus, even while Britain was trying to police slave vessels across the Indian Ocean, the trade continued and the slavers continued to prosper.

Mackinnon, who had tried to construct a useable road from the coast to the centre ('the Mackinnon road') had written to Lord Salisbury, the then Prime Minister, a letter which concluded 'with [some financial] support, the Company (IBEA) believe that slave-trading would soon disappear' with the construction of a railway – Mackinnon's goal.[5] Lord Salisbury was convinced, stating in a speech in 1891 that:

> Its [the railway's] object has been... to deal a deadly blow at the slave trade, the destruction of which has been, along with our own commercial and material progress, an animating impulse in English policy in those regions....The slave trade on the sea only exists in the eastern coast of Africa and on the shores of the Red Sea... but the place

> where caravans still go... is the tract that lies between the great Victoria Nyanza lake and... Mombasa.[6]

Public disgust at the slave trade was politically potent.

The argument ran that, once built, a railway would out-compete every other type of transport, and would certainly kill off the caravans which used slaves to carry ivory to the coast. It would cost two or three hundred times as much to bring goods by caravan as to bring them by railway, so that once a railway existed, the caravans would become 'a matter of antiquity'. In other words, without caravans there would be no means of bringing slaves from the interior to the coast 'because', Salisbury continued, 'I do not see that any slave trader who presented himself with a body of slaves to be carried on trucks to the coast would be very civilly received'. The economic benefits of the railway were rapidly understood by the caravan traders, for as soon as the railway had crossed the dreadful Taro Desert, the traders used it to speed up their journey which proved more profitable than using the slow transport mode of slaves and head loads. The railway reduced the journey time from Mombasa to Nairobi from six weeks to 24 hours.

At the same time, there were strong strategic reasons for a railway to link Central Africa – in particular Uganda – to the coast. The European powers had been scrabbling over the headwaters of the Nile for some time – and France and Britain were the main protagonists in this tussle. One particularly intriguing confrontation between the British and the French took place at an abandoned mud brick fort at Fashoda on the Upper Nile, seen to be a key strategic point. This was a confrontation which almost took France and Britain to war as the two countries raced to claim it. The British approached Fashoda equipped with the military might of Kitchener's gunboats while, in a highly secret mission, a handful of French officers and a group of Senegalese soldiers had slowly made their way on foot and by river from the Congo in the west. They had brought with them a portable steam launch which could be dismantled where there was no navigable waterway, and which they carried, pushed or dragged across the country as the terrain demanded. They also brought plentiful supplies of champagne and claret, a mechanical piano, and seeds of *haricots verts* to plant in the Nile mud. When the British arrived they found well-tended gardens and good cuisine, but although the French had arrived first, the British won the encounter (by some perfidy). The tale serves to illustrate the lengths to which these competitive European states would go.[7]

Elspeth Huxley expresses the logic of this scramble and the strategic significance of the railway with eloquent and direct economy. 'The strategic reason was... built up as follows: whoever rules Uganda controls the Nile; whoever controls the Nile, dominates Egypt; whoever dominates Egypt holds the Suez Canal; and whoever holds the Suez Canal has his hand upon the throat of the India trade'.[8] A railway across Kenya to Lake Victoria would provide the means to secure Uganda.

Amid great controversy in London, funds were eventually voted in Parliament in 1895 for the construction of a railway from Mombasa to Uganda. The main reason the vote succeeded in the end was the control and elimination of the slave trade. There was no thought at the time of trade and development since there was little to trade and no one to buy British produce. It seems likely that Realpolitik and the strategic reasons were at least as important as the desire to rid East Africa of slavery, since there were no significant slave caravan routes from Mombasa. In fact, the railway rapidly became very useful for transporting people, goods and troops; in 1898, when about 100 miles of track had been laid, troops and equipment were carried by train to the railhead to subdue a rebellion in Uganda, and this stretch[9] was then opened to public goods traffic in February of that year.

The construction of the 'Lunatic Express' itself is a story full of bravery, danger, horror and achievement.[10] The labour force consisted mainly of Indians who were employed on contract – 'coolies' – and supervised by one George Whitehouse who had been appointed as Chief Engineer, charged with the construction of a railway from Mombasa to Lake Victoria and Uganda. An engineer with experience of building railways in South America, India, Mexico and South Africa, Whitehouse was nonetheless daunted by this prospect. The route had been given a preliminary survey three years previously, but was by no means fully surveyed; and while to some extent he could follow the trails of the caravans, a part of the route went through the Taro Desert west of Mombasa, a barren and bleak waterless region that the caravans tended to avoid. Beyond, the route forced its way through some three hundred miles of gradually rising savannah and scrub country that teemed with lions, tsetse flies and malaria. After this 'came a ragged volcanic highland region, which was bisected, as if with a cosmic meat cleaver, by the fifty mile wide Rift Valley',[11] the great trench which extends from Syria in the north to Mozambique in southeast Africa. While explorers had described the Rift as one of the great areas of natural beauty, to Whitehouse it was a huge engineering challenge requiring an engineering miracle; the manhandling of rails over the valley's near vertical, 2,000-foot escarpments. Beyond lay a further hundred miles to the lake which was thought to consist of 'a sponge-like quagmire'.

Whitehouse had been given four years to reach Lake Victoria, but, by December 1896, at the end of the first year, only 23 miles had been laid, from Mombasa to Maria Kani. Problems encountered included rains so heavy – nearly 30 inches in November – that plate laying was delayed by three weeks and the new earthworks suffered from subsidence; and a quarter of the workforce was hospitalised with dysentery, malaria and various jungle sores. Labour disputes in England delayed the production and delivery of rolling stock. The first 20 miles out of Mombasa turned out to be an area very heavily infested with tsetse fly, and 1,500 out of 1,800 transport animals were killed by tripanosomiasis. As Mervyn Hill says, 'From first to last the building of the Uganda railway was a tale of hopes deferred, of disappointments, of estimates unfulfilled, of coping with unpredicted and often unpredictable difficulties'.[12]

The most famous of his trials was the encounter with the man-eating lions of Tsavo. Apparently lions rarely eat humans, and indeed dislike human flesh, but in this particular area, man-eating lions held up works on the construction of a bridge over the Tsavo River for ten months and killed and ate 28 workers.

> [The] work was interrupted in a rude and startling manner. Two most voracious and insatiable man-eating lions appeared on the scene, and for over nine months waged an intermittent warfare against the railway and all those connected in the vicinity of Tsavo. This culminated in a perfect reign of terror in December 1898, when they succeeded in bringing the railway works to a complete stand still for about three weeks... as time went on, they stopped at nothing and indeed braved any danger to obtain their favourite food. Their methods then became so uncanny and their man stalking so well timed and so certain of success, that the workmen firmly believed that they were not real animals at all but devils in lions' shape.[13]

Eventually this pair was killed but problems from lions continued, and their audacity was astonishing. Three months later, an engineer, O'Hara, was sleeping in a tent with his wife and children beside him. While they slept, a lion killed O'Hara instantly by crushing his head in its jaws, and then dragging his body from the tent without waking his wife. At another stop, at Kima, the Superintendent of Railway Police, Charles Ryall, undertook a vigil to trap a particular lion known to prowl in that area. He was accompanied by two others, with the idea that they would take turns to sleep while one kept watch, with the aim of shooting the lion when it emerged. Alas for Ryall, he fell asleep when he was meant to be on watch, and the lion seized the opportunity, strode directly into the carriage, broke Ryall's neck with a swipe of its paw and leapt from the carriage with the corpse between its jaws.

In spite of these problems, by June 1899, the railway had reached Nairobi at 5,500 feet with 325 miles constructed. To reach here, they had endured drought, famine and smallpox. Nairobi was a pretty bleak and unprepossessing place in those days – little more than a shanty town – but it was the last reasonably level ground before the land rose steadily for a further

2,000 feet over 27 miles to the summit of the Kikuyu escarpment. It was marshy – the name Nairobi derived from the Masai name Ewaso Nairop, meaning stream of cold water – and largely empty but for swamp and frogs. But it served a purpose. By the end of 1899 the railway headquarters were established there, and, by the following August, a more or less regular passenger service between Mombasa and Nairobi was operating, and the town began to take shape. A main thoroughfare was built, called Victoria Street, although it became a quagmire in the rain. There was a collection of Indian *dukas*, or general stores, and an Indian bazaar. There was a ramshackle wooden hotel which doubled as a general store and a corrugated iron building which became the headquarters of the Railway Members Club – and the haunt of the other Europeans. Housing consisted of tents and corrugated iron bungalows but an embryonic municipal council was set up with Ainsworth now appointed as District Officer for Nairobi. Over time Ainsworth would transform Nairobi with the planting of exotic eucalypts and jacarandas.

The rest of the journey was not straightforward, but by 1903, after clambering down and up the sides of the Rift Valley, recovering stolen cable and machinery from the Nandi tribe and several more requests for money from London, the railway reached Lake Victoria at Port Florence – the port named after the wife of the chief engineer (later it was renamed Kisumu). The first stage, 572 miles, of the lunatic express was complete. Since 1986 when construction had begun the total cost had been £5.2 million, more than twice the optimistic £1.9 million 'final request' which had originally been submitted to the London government – showing that cost overruns on major projects are not just a phenomenon of the twentieth century, although at £9,000 a mile, roughly equivalent to £1 million per mile in today's prices, it was not dear when compared to some modern railways.[14] But the way now lay open to the settlement of East Africa.

The railway runs through 500 miles of (then) economically undeveloped land, and although the first 300 miles of this – from Mombasa to Nairobi – were largely waterless desert, the land between the Kikuyu escarpment and Lake Victoria was potentially rich, and it was here that development was planned, settlement was then logical and inevitable. The initial investment cost was borne by the British government, but this was not a long-term option – the railway had to be made to pay. The subsistence economy of the African did not seem to offer enough potential for trade and surplus; the general practice among the cultivators (not the pastoralists) was a shifting agriculture, where land would be cleared and cultivated for a few years until the soil fertility became exhausted, whereupon the community would move on to clear a new patch of land for cultivation. Even among the pastoralists, little was traded and head of cattle represented wealth and status rather than cash riches. To the government of the day the higher reaches of the railway passed through lands which appeared both empty and fertile. It seemed to the government that the only feasible way to develop a trade along the railway was to encourage settlement by immigrant farmers, who would produce a surplus and realise the potential of this apparently wonderful, empty, country.

And with the railway and the first elements of a city, British East Africa was attracting, as frontiers usually do, a rich mixture of men and women from many countries and all walks of life – missionaries and merchants, priests and planters, big game hunters and brick makers, builders and bankers, developers and con men, surveyors and scientists, settlers and explorers, aristocrats and artisans. The one thing they held in common was hope that: 'in this new land they would find something that they had not found at home'.[15]

Apart from a few early hunters, the first to come in earnest were the missionaries. Missionary activity in East Africa started in the nineteenth century and was strongly linked to the commitment to the abolition of slavery. The earliest had been Johann Krapf, a linguist, explorer and theologian who was seen as the founding father of the Anglican Church of Kenya, who arrived in East Africa in the mid-nineteenth century. Thereafter there was a steady trickle of missionaries; records show that from the Church

Missionary Society (CMS) alone about 150 missionaries, one third of them women, came to settle in East Africa during the latter part of the nineteenth century and the period up to the First World War. In 1887 the *Church Missionary Intelligencer* carried advertisements: 'Wanted immediately: three ladies for East Africa. Must be whole hearted missionaries, physically strong and thoroughly understanding the principle, "In honour preferring one another".'

Between 1900 and 1913, the CMS mission set up seven stations throughout Kikuyu land. There were also missions established by the Church of Scotland, the Africa Inland Mission and the Gospel Missionary Society. Thus, by 1915, there were four separate Protestant missions, and a total of 15 stations in Kikuyu land alone. They established clinics and mission schools, and linked conversion and redemption to education and work. The government actually funded educational missionaries. They were also brave and pioneering, undergoing considerable difficulties in the belief in the rightness of their mission.[16]

The diary of Mary Grimes, who arrived in Kenya in 1905, describes graphically her journeys with her husband to set up a medical mission in Meru, on the lower slopes of Mount Kenya, and shows the tenacity and conviction of these people. Dressed in the corsets and multiple layers of the late Victorian period, she was carried in a sedan chair over mountains and through streams by native bearers. A photograph shows her in a kind of deck chair carried on poles by four Kikuyu men, naked but for leather aprons. She writes of her journey through dense bamboo forest over the slopes of Mount Kinangop, a journey of 60 miles which she undertook accompanied by the American head of the regional mission station. She set off with her escorts at 7.30 in the morning, carried in her sedan by the four Kikuyu and accompanied by three other missionaries. She writes how:

> we... began climbing a tremendously steep hill where my chair had to be dragged through dense undergrowth, while at the same time we were impeded by overhanging branches... the remainder of that day was spent traversing an enormous plateau with Mt Kinangop looming dark

> before us... towards the end of the day we neared the bamboo forest and there was Mt Kinangop waiting to be scaled... after breakfast of boiled rice... we were on the march again and the path became very overgrown.[17]

She heard that only one European lady had ever been through the bamboo forest before, the wife of the Fort Hall official. In parts very swampy, her bearers were frequently struggling with deep mud, and it would require all the additional porters to steady her chair. Later, the chair had to be hauled up almost inaccessible places, dragged through a near impenetrable wall of bamboo. She notes that it was worth a great deal to see the beauty of the forest with its trailing vines, begonias and mosses; with plenty of long-haired Colobus monkeys in the trees above.

But the descent was in drenching rain, with mountain torrents to cross. At one point in the journey, they had descended an almost vertical hill in continuing rain, 'the rain poured persistently down the back of my chair so that I was soon sitting in a pool of water'; and came upon a magnificent waterfall and, just above it, the Maragua River, to be crossed. In her voice:

> A more impossible place to carry a lady in a chair could scarcely be found! Yet, being but a poor climber I could not trust myself on foot; and so it transpired that I was borne down over the rocks and across a river in a manner that was truly incredible... the streams were bridged only by a single plank and indeed the term plank is altogether too civilised a term to use, as in each case it was simply the trunk of a tree, roughly hewn in half, and rendered very slippery by mud and rain. The second of these was of great length and some eighteen feet beneath it was a rushing stream with sharp stones jutting out of its rocky bed.... I dared not venture across such a narrow and slippery bridge on foot, and, as it must needs be traversed, I committed myself to God, and remained in my chair. Had a single porter slipped we must have been hurled into the river, the balance was so extremely critical; but God mercifully kept their feet and brought us in safety to the other side.[18]

Once arrived at her destination, she lived under canvas for some weeks in the autumnal rains until the corrugated iron house was completed. The commitment of the missionaries was extraordinary, the burning need to bring Christianity to the heathen, however remote, and over time they became the champions of the indigenous peoples when tensions arose with settlers and government. The following passage shows clearly the depth of belief that they were doing the right thing:

> One is not long in the mission field without realising that there is a hard spiritual fight to be fought.... Our teaching naturally strikes at the very root of the foolish superstitions and degrading customs with which the Prince of Darkness seeks to bind these primitive people body and soul. In the clear light of gospel truth they stand revealed in their true character, and the pupil at the mission school learns to regard them with aversion. 'The entrance of Thy Word giveth light'. He can no longer bear to participate in what he now sees as displeasing to God. From that point of course the real conflict begins. *But for those who know that they are on the winning side there is nothing to fear, tho' temporary rebuffs may be encountered. The 'Captain of their salvation' has never known defeat, and the weakest of His soldiers may learn to triumph in his blessed Name* [author's emphasis].[19]

Her memoirs show the link between missionary work and the abolition of slavery:

> Ah! If the grand old mountain could only speak what stories it could tell us, not only of the raids of the Masai but of the coming of the Arab caravans, of the remorseless cruelty of the slave dealers, the pitiless lash and thong, and the unavailing cries of those who were torn away from their homes and kindred never more to behold them. Alas![20]

And she goes on to say, with a wry self-knowledge, 'But with the coming of the British dawned a somewhat brighter day for the

Kikuyu tribe, though they were naturally slow to recognise the advantage'. But social change was inevitable and began to loosen some of the traditional structures of society. So, while she and others made converts among the Kikuyu, once a young man had become a Christian, he might refuse to participate in traditional sacrifices – described by the missionaries as heathen practices – and thus the bonds among the Kikuyu were weakened, because many Kikuyu ceremonies and rituals of religious worship required all male members of their family to be present for their validity.[21]

With their commitment to learning local languages and their close involvement with local communities, the missionary community were a powerful force for change – much of it good – in East Africa. They also believed in work and industry, tying it in with reading and writing.

Mary Grimes expresses these ideas thus:

> I must not forget to mention the industrial side of the work. We felt very strongly that to develop the African along the right lines he must be taught to work with his hands, while at the same time learning to read and write. For this reason, although numbers of boys lived on the station, no money was ever paid out to them for their support, but only for actual work accomplished. On this account we endeavoured to find employment for as many as possible. Brick making, house building, rough carpentering and gardening, beside the mat industry, kept a large number of men, women and boys constantly busy for hours a day. Returning from school each morning I was generally followed by a crowd of lads begging for work; and two or three tiny mites of four or five years of age would press up to me flourishing their little cultivating knives aloft and crying lustily 'Hee kawera, hee kawera' – Give me a little bit of work.[22]

Thus, by the time the first farming settlers came, there were already four or five different missions[23] and perhaps 30 or 40 mission stations established in Kenya. Missionaries opened schools and taught trades as we see from Mary Grimes' memoirs.

They translated the Bible and the liturgy into local languages. They were the foundation of education for Africans and the mission schools were supported by the government. With their considerable knowledge of local communities and beliefs, and despite later bitterness over some tribal practices, notably female circumcision, an issue which caused a severe rift between some Kikuyu groups and the mission schools, the missionaries were often able to present the voice of the African in politics and government.

Among the best known of the missionaries in the early twentieth century are Archdeacon Walter Owen and Canon Leakey. Archdeacon Owen was born in 1879 in Birmingham and later trained as a missionary in the CMS training centre in Islington, London. He arrived in Africa – in Uganda – in 1904 and in 1918 was moved to Kavirondo – then still part of Uganda. There he stayed fighting the government on issues such as forced labour and supporting African rights in the settler dominated politics of Kenya. In 1921 there was an uprising among the Kikuyu led by Harry Thuku, a telephone operator who set up the Kikuyu Central Association (KCA). Archdeacon Owen worked to convert the uprising into a more moderate, less political, organisation and set up the Kavirondo Taxpayers' Welfare Association in 1923, which worked to bring welfare and education to the African population and encouraged the members to plant trees, dig sanitary pits and provide themselves with a bed. Archdeacon Owen was supportive, for example, of the desire to grow coffee by Kikuyu farmers. He was continually active in the question of land, fiercely denouncing settlers and gold prospectors alike for expropriating African land – he even suggested that missionaries should pay compensation to those Africans whose lands they had used for building the early missions. He was a forthright activist for African rights at all levels.

Harry Leakey had arrived later, in 1901, and settled in Kabete, Kikuyu. His daughter Gladys later married Leonard Beecher, Owen's school contemporary, and his son Louis became one of the most famous palaeontologists. The Leakeys all spoke Kikuyu fluently, and the extent to which missionaries shared and understood the lives of Africans is illustrated in Louis' book *Mau Mau and the Kikuyu* where he wrote:

> I was born and bred among the Kikuyu people, where my parents, who were CMS missionaries, lived and worked for so long. I speak the Kikuyu language as well as, if not better than English. I spent most of my life with the Kikuyu, I am a member of the Kikuyu age group called *Mukanda*.... I am also an initiated first grade elder.[24]

Missionaries like these saw their roles as protecting the rights of Africans and of course spreading Christianity – within the colonial structure. They would not have been revolutionaries against the colonial system – it was too well respected in those days; the belief in the virtue of empire and the civilising mission was very strong, to an extent which seems improbable to later historians who unequivocally see empire as wrong.

After the missionaries came the settlers: the farmers who would generate the surplus to provide freight for the railway, although it took longer to attract them in any numbers. The potential for European settlement had first been identified by Lord Lugard who was travelling on military duty in East Africa in 1893. The Mau was empty of any settlement when he crossed it, and the Uasin Gishu and Laikipia plains were also practically unoccupied. He thought the Mau was a good place to start because it was uninhabited and 'the soil is extremely rich and covered with excellent and luxuriant pasture throughout the year with which is mixed white trefoil and clover' and he imagined the development of fruit crops and cattle – following the examples of New Zealand and California. But settlement was not formally encouraged until the railway had been built.

Over the following five years the government struggled to establish appropriate land ordnances – ones that would encourage settlers to come and invest, without their being bound by excessive bureaucracy; but that would protect the native population and their traditional farming and grazing lands. By

1902 a second Crown Lands Ordnance was passed in order to control the disposal of land, and under this all land, whether occupied or unoccupied, was deemed to be Crown Land. The land allocated for white farmers in the first instance was on the high plateau of Kenya and around the west and south of Mount Kenya – an area which came to be the so-called 'White Highlands' – and although immigration was actively sought, there was never a deliberate intention of displacing native peoples – at least by the government in Whitehall. The areas which were not to be settled by immigrant Europeans were designated as Reserves and were reserved for the indigenous population, a policy which was largely sustained. But Sir Charles Eliot,[25] commissioner for the territory at the time, felt free to turn the empty space along the railway line over to 'settlers who would turn the fertile but now wasted soil to useful account, who would grow crops for the railway to carry and buy machinery and other goods for it to carry in'.[26] He identified strongly with the needs of the pioneer settlers and tried – ultimately without success – to ensure that there was no restriction in the area of individual holdings. Eliot's vision matched precisely the aims of the colonialists: settlement and the realisation of the value of the land.

And, despite the aims of Whitehall, alienation of land for white farmers caused lasting and deep resentment among African communities, particularly the Kikuyu. In 1904, about 60,000 acres of Kikuyu land in Ukamba province at Kiambu and Dagoretti were alienated by the British for settlement. The area was sparsely populated since towards the end of the nineteenth century a series of disasters – famine, smallpox, rinderpest and locusts – had pushed the Kikuyu away from some of their land, and as a result this land appeared – wrongly – to be unoccupied and available. This error had profound and far reaching effects. The Masai too were moved from traditional grazing lands – to other lands of the same area – but nonetheless the process and actuality caused a continuing discontent.

The first real wave of settlers arrived in 1903 and 1904.[27] One of the first pioneering land owners was Lord Delamere. Born

Hugh Cholmondeley, son of the second Baron Delamere, he grew up on a fine estate, Vale Royal, in Cheshire, in an essentially feudal world, a background that was in his bones and blood, and he believed all his life in its fundamental rightness. Hunting was his passion, and it was not long before he ventured to fashionable East Africa as a big game hunter, in the early 1890s – following a route through the Horn of Africa, a journey which took him to Abyssinia, across harsh and barren deserts, until he finally arrived in central Kenya – on the Laikipia plateau. According to Huxley, on whom we rely for her biography of the great man, 'here, (he must have thought) is an Eldorado... a country of great latent wealth only waiting for development... a friendly and temperate patch of Africa where a white man could feel alive and invigorated, where he could keep healthy'. He did not settle immediately, but returned to Kenya in 1899 with his new wife, Lady Florence Cole, again to collect specimens, and then, encouraged by Sir Charles Eliot, Delamere finally settled in Kenya, captivated by the landscape and its apparent potential.

Delamere saw in himself a pioneer who could invest and inspire people in this country of great promise, and who could 'show that in East Africa England had possessed herself of a little New Zealand tucked away between deserts and tropics and lakes, where yet another cutting from the British parent stock could be planted and grow and flourish'. He believed in the Highlands as a 'white man's country', a vision he and Eliot shared:

> If we administer the country with foresight and rectitude... and execute the dictates of common sense without muddling, few who know the country can doubt that it will shortly be a flourishing European Colony... and its development will mean the opening of a new world.[28]

He obtained 100,000 acres on a 99-year lease for which he was required to pay £200 annual rent and to spend £5,000 on the land over the first five years of occupancy; this estate he called Equator Ranch.

The land was spectacular. Huxley writes:

> As far as the eye could see there was no sign of human life. There was nothing to soften the immensity or to give a sense of proportion to the landscape. Just two huts [where he and his wife set up home] with no garden around them, not even a fence or a hedge to divide the homestead from the veld. Oxen could wander up to the door and look in, wild birds investigate. Everywhere colours were vivid and deep and always changing, distant escarpments sharply defined the sky rich with moving clouds, the atmosphere clear with the rarity of the mountain air and charged with the sense of space and freedom blown on cool, crisp breezes from the hills.[29]

Inspired by his vision, he saw the great wide plains below the Mau, near Njoro, as sheep ranching country and in pursuit of this he poured thousands of pounds into different projects: he brought cattle and a farm manager from Scotland and 500 merino ewes from New Zealand; he improved the pastures and interbred local cattle with Herefords; he tried dairy farming. But the cattle wasted away – the land was deficient in iron, or they were killed by redwater fever. The sheep were destroyed by tick fever and other unknown viruses. He moved onto dairy farming – but he had not reckoned with East Coast fever – a microscopic blood parasite that killed off all young stock in infected areas. He spent a vast amount of money on these schemes, some said as much as £80,000, mostly borrowed. And after the livestock had failed, he went on to wheat farming, breeding seeds to resist rust, and later established the first dairy industry in Kenya. This willingness to experiment, follow paths of trial and error, showed a tremendous tenacity, and one which would ultimately transform the land of the White Highlands.

Delamere was an extraordinary character, usually in scruffy clothes, with a mane of red hair which he wore down to his shoulders; he fought regularly with his neighbours, was notorious for creating mayhem in bars and hotels – he once rode his horse onto the bar at the Norfolk Hotel – and had a ferocious temper; on another occasion he locked the manager of the Norfolk Hotel in the meat safe with several dead sheep because he was irritated by some constraint on his behaviour. Yet he had a reputation as a careful and committed farmer, and built a good relationship with the Masai. But his commitment was to white farming and white settlement; his formidable energy was focused only on this and he – as the vociferous leader of the settlers, as he became – stood equally for hostility to the Indian community and to the development of African agriculture, determined all the while to establish a white colony in East Africa and to limit absolutely the land available to others. A fine bronze statue of him was erected in the centre of Nairobi, at the head of Delamere Avenue – this was moved following Independence to the site of his former ranch and the street name changed to Kenyatta Avenue.

Many other settlers also came from an aristocratic background, to the extent that the Norfolk Hotel was sometimes referred to as the House of Lords. One was Cecil Hoey, Delamere's brother-in-law who had arrived in 1902 with wagons, which he dismantled and carried by hand through the Aberdare mountain range. There was also a strong contingent of South Africans and one of the more famous groups to arrive was the great Boer trek to the Uasin Gishu in 1908. Hoey described their arrival: 'I was sitting one day on top of a rock called Sergoit to the west of the Elgeo Escarpment watching three lions at play, when I saw on the horizon a long white streak which at first I thought was smoke'. Then it became too definite a shape for smoke and it looked more like a river, though this could not be. Gradually the line grew closer, and he was able to make out through his binoculars a long line of ox-drawn wagons, their covered tops showing as a dusty white. They had arrived from the Transvaal, by boat across the lakes and a final long trek overland, a group of 47 families, three men and two predicants, led by one Jansen van Rensburg. They leased land from the government on the Uasin Gishu plain. At the time of their arrival, the land was empty of African settlements, 'a great plain of waving grasses… trees and rivers and vast herds of game' – this they thought was as the Transvaal had once been and was no more. Twenty-five years later the plateau was transformed into wheat and maize fields, fenced pasture with cattle bred from local zebu and pedigree European stock, flocks of sheep, fish ponds and plantations of black wattle tree.[30] But no longer any lions, nor other game.

The South Africans had a strong impact on settler attitudes, reinforcing the beliefs of those who, like Lord Delamere, saw Kenya as a new white colony, like South Africa and New Zealand. But people came from many walks of life, and for many different reasons – some came to farm, but others came to work on transport or business. They came from many corners of the world. Britain and South Africa – but also from Australia, New Zealand, Scandinavia, France, Ireland and India; attracted by new opportunities and adventure. Some jumped ship from the Merchant Navy, or were fleeing justice from other dominions. Others were sent as remittance men, wayward sons of respectable families (indeed, Mervyn Hill was sent to Kenya, ostensibly to cure his asthma but also to get him away from his family[31]). Others from Canada or Australia migrated north from South Africa after fighting with the British forces in the Boer War, arriving in Kenya as railway engineers or traders and stayed, perhaps looking speculatively for land. Kenya attracted an energetic and pioneering population, women as well as men.

Besides the European settlers, there was a steady stream of Indian immigration – and migration from the coast inland. About 32,000 Indian coolies (and their families) were brought from the Punjab for the construction of the railway. Most of them were on short-term contracts, and the majority left when the railway was complete, but about 6,000 elected to remain (the Indian government had stipulated that they should have the right to stay if they wanted to) of whom about one-third continued to work with the railway.

There was already a historical presence of Indians in the coastal areas of Kenya. Indians came from Gujarat as financiers and administrators to the sultans of Zanzibar and there is evidence of trade between India and the east coast of Africa going back 2,500 years, since the coconut palms which characterise the coast of Kenya came on the Monsoon Trade from the Malabar coast. The Indian traders and financiers were well established in Kenya, and as the railway was completed, the older established Indians were encouraged by the British to move inland.[32] The influence of Indians is attested to by the fact that the rupee was the currency in Kenya until 1920. By 1898 Indians were trading in Machakos and at the missionary settlement of Kibwezi. There were a number of Indians who amassed fortunes as railway contractors and builders, such as A. M. Jevanjee, the head of a firm of contractors, ship owners and general merchants and the first Indian to be represented in the colonial government with a place on the Legislative Council.[33] He had arrived in Nairobi in 1899, to build the residence of John Ainsworth, the first commissioner for Nairobi, on Museum Hill and became a very prominent builder and developer in Nairobi. In 1902 he established the first English language newspaper, the *African Standard*, a weekly, published first in Mombasa and later as a daily in Nairobi. In 1906 Jevanjee presented a statue of Queen Victoria to the town, which was unveiled in the Jevanjee Gardens by the Duke and Duchess of Connaught.

Another great Indian trader was Allidina Visram, an Ismaili who came to Africa as a boy in 1863 in the caravan trade and specialised in ivory (he was known as 'King of Ivory'). His business career was extraordinary; he banked the pay of the Indian navvies in the 1890s, sailed a merchant navy on Lake Victoria to bring cotton from his ginneries in Entebbe, Uganda; grew rubber, ginned cotton, refined sugar and sold insurance in Uganda; he boiled soap in Mombasa and imported bicycles, owned 200 shops, and had built a commercial empire throughout East Africa and into the Congo before he died in 1916.[34] But in the early days Indians were denied land rights and the main option available was to set up as petty traders, creating a network of *dukas*, which became the backbone of communication in Kenya. The settlers fought hard for many years against any political recognition or rights for the Indian population.

There were two immediate and far reaching consequences of the final dominance by Britain of Kenya. The first was the so called *pax Britannica*, the efforts to reduce fighting between warring tribes whose pattern of existence had since time immemorial included cattle raiding by the Masai on the Kikuyu and battles between, for example, the Gusii, a Bantu people, and the Nilotic Luo. One of the strategies of land alienation was to put splinters of European farmed land between warring tribes to eliminate fighting and raiding. Thus what became the pyrethrum and wheat farms in the Uasin Gishu effectively separated the Elgeyo to the east from the Nandi to the west. The white settlement at Sotik in eastern Nyanza Province was set up as a buffer zone to stop raiding between the Kikuyu and the Kipsigis. The changing pattern of inter-tribal conflict and ritual was to have profound impacts on the social and political structures of the country. It released young men to work on European farms, but they had been trained as warriors – farm work was a poor alternative to the glory of fighting.

The second, even more transforming of tribal society, was the introduction of the Hut Tax in 1902. The Hut Tax, levied as its name suggests on the huts inhabited by the native population, was a standard British way of raising taxes from the native population, in regions where there was no formal economy. Around the turn of the century the tax was Rs3 per hut per year[35] or about £10 in modern day terms, but it rose to Rs5 after the war and then to Rs10 (20/-) – although this level was soon reduced to Rs8. The Hut Tax was introduced on the one hand to raise revenues for public works by the government (roads in particular) but it was also a mechanism which would push subsistence farmers into the labour market, an outcome which the white farming settlers viewed as nothing less than their right. Until the advent of the Hut Tax, the African population was reluctant to undertake paid work on European farms; they had no need to find cash on a regular basis, living as they did a comfortable subsistence way of life with trading as a side activity and accumulation and wealth measured in cattle and goats. Now, however, Africans had to find cash on a regular basis, to pay the tax, and in many areas of Kenya this was the first time the people had encountered wage labour – or the need to exchange animals for cash. Money had been introduced into East Africa by the

IBEA but had little impact on the lives of most of the population. But the Hut Tax achieved its goal of pushing people to paid work on settler farms and the settlers naturally supported the Hut Tax vociferously. Once cash was required, its use spread; traders began to want payments in cash rather than goods for example. Now wealth could be measured in rupees rather than in cattle, the security and rituals of exchange became more fluid – one might say that this is the beginning of civilisation, but it took time for the negative repercussions to work through society, and when cash came to replace traditional exchange, the relationship between men, women, and work was changed entirely.[36]

Another source of labour was provided through the squatter system which was welcomed by settlers and African alike. Typically, the African would provide 180 days of labour to the farmer, but could bring their own families and could cultivate their own *shamba* on the settler's land. The squatters were mainly Kikuyu who gained access to land and provided labour for the farms, often in areas which had not been previously penetrated by Kikuyu. As well as providing new land, the system enabled Africans to develop new farming skills. From the employer's point of view, if they had plenty of land, the system provided a constant supply of labour as needed, which did not need to be laid off in quiet seasons as the squatters had their own patches to cultivate. Squatters worked under three-year contracts, but these were frequently renewed. Government departments, mainly the Forestry Department, employed large numbers of squatters.

At worst Africans on the settler farms were treated with brutal cruelty. In early years the rhinoceros whip was widely used. One early settler, Arnold Paice,[37] catalogued in letters the punishments he inflicted on his African 'boys' – punishments which ranged from 'a pretty good thumping' to shackling them by the enlarged lobes of their ears. Beating of servants was common. Lady Cranworth in 1912, in a textbook for new arrivals commented:

> supervision of native servants is an art in itself. One could not, for instance, learn by experience in England when is

> the right time to have a servant beaten for rubbing silver plate on the gravel path to clean it, and this after several previous warnings.[38]

But this is only one side of the settler story. Many of the men wrought good relationships with their African workers and managers, and many of the women set up schools and small infirmaries for the African squatters and their families. There were many examples of good humane treatment. In writing of her parents' farm Elspeth Huxley describes a generally decent way of treating African labour, with the understanding that access to land worked for both farmer and squatter. After the First World War when her parents moved to Njoro, she writes of how, when her mother Nellie arrived at Njoro station (having had to leave the coffee estate at Thika), she found waiting at the station half a dozen Kikuyu employees who had gone from Thika to accompany her to the new land.

> She was touched at this act of loyalty and affection, but knew that loyalty and affection were not the primary reasons. Word had got around that there was fresh land, with plenty of firewood going free as well as grazing for cattle and goats.... When it came to giving out new *shambas* they wanted to be in on the ground floor. Now (ten years later) clusters of round thatched huts... had arisen. When one Kikuyu family comes, others follow. The half dozen individuals she had found waiting at the station had mushroomed to a sizeable population, each family with its goats and cattle and its children.... Already Nellie had started a farm school.[39]

A further example was Olga Watkins who amazed her husband-to-be Oscar, an administrator, when she first met him with the quality of her farm, the properly built houses for the workforce and the school for the children. Oscar himself was a committed defender of African rights, and, like many others in the administration, had no great liking for the aristocratic settlers who made Kenya their playground, and felt strongly that the role of the administrators was to protect the local inhabitant from having his land occupied or his labour exploited.

From a European perspective settlement was about civilisation, the zeitgeist of the age; civilisation, the inherent virtue of work and a great confidence in the appropriateness and superiority of European behaviour – combined with the search for raw materials for the great industrialisation process. The great energy and vitality of the industrialisation and economic expansion, the ability to make the most of technological breakthroughs and new mechanisms for accumulation were – perhaps hubristically – seen as essential virtues at the time. The nineteenth century had been an age of boundless optimism – and it carried with it a great concomitant sense of philanthropy. It is unlikely that the drives to end the slave trade would have emerged without the growing wealth and education in northern Europe. The great philanthropists and benefactors of the Victorian age combined with missionary activity and the expansion of the missionary world, in a civilising enterprise which sought, for example, to ensure that all peoples could participate in the modern world through education and Christianity – but also required that they covered their modesty and wore clothes, or that young men should turn to economic activity rather than tribal warfare.[40] These assumptions had wide and deep impacts for the 'beneficiaries', but were in the mainstream of nineteenth century thought.[41]

The settlers brought enterprise and vision, often taking great risks in their investments. The government brought the rule of law and a commitment to development of the indigenous population – over time of course. And between them, society in Kenya was transformed irrevocably. One has the impression of Kenya as a lively, vibrant place by the mid-1920s. Immigrants were coming from a more diversified background, and the settlers brought with them energy and a readiness to take risks and try new things. The Asian population was effective in providing commerce, delivering goods and services throughout the country, and brought shops and markets to settlers and Africans alike. On the other hand the vision of development and trusteeship attracted a caste of colonial officers who saw work within the government and colonial service as service and worked with commitment to these ends. The two were in some sense reflections of each other, opposite sides of the same coin; the energy of the settlers enabled the early establishment of a robust economy in Kenya, while the administration provided the legal governance and controls necessary to protect and safeguard this new economy. But in many ways, settlers and officials were at loggerheads from early on – with the native population squeezed between the irreconcilable motives of each. Many of the early administrators – John Ainsworth, for example, and Oscar Watkins – took the notions of trusteeship very seriously and saw as their mission the protection and development of the indigenous population, and in this they were supported by the commitments of the missionaries. But the settlers wanted a free hand to build their own future untrammelled by notions of trusteeship or controls and over the decades the governors of Kenya would be seduced and cajoled by the settlers to give them all the freedoms they wanted – at the expense of the African, and Asian, population.

So when the young agriculture officers arrived, full of hope and a vision of service, they arrived to a rapidly changing society. The role of the young agricultural officers who were assigned to native agriculture was to work with the indigenous population in the Reserves to modernise and improve agriculture along the lines set out by Alex Holm. But it was far from a *tabula rasa* by then and the pattern of early settlement was to underlie much future suspicion which would bedevil their work, at the heart of which lay, always, the question of land.

3

OFF TO NAIROBI 1931–34

'Nairobi was set to be the metropolis of East Africa!'
Mervyn Hill, 1949

'Maybe one day Nairobi will be laid out with avenues of flowering trees, flanked by noble buildings; with open spaces and stately squares; a cathedral worthy of faith and country; museums and galleries of art; theatres and public offices. But [for now] Nairobi must remain what she [is], a slatternly creature, unfit to rule over so lovely a country'.
E. A. T. Dutton, *Kenya Mountain*, 1929

The Nairobi to which Owen moved in 1930 was raffish, squalid and dirty. Margery Perham, arriving after a long journey through other parts of Africa in 1931 writes:

> Nairobi itself is a most disappointing town. I had expected something rather smart and well built. It is one of the shoddiest and shabbiest towns I have seen in my travels, which is saying a great deal. There are hardly any pavements and the roadway itself is most primitive. You either stumble through mud or are blinded by gritty dust. The only decent building is the railway office. (Government House is outside of the town). Most government offices are tumbledown tin shacks; the Supreme Court is like an abandoned warehouse.... The New Stanley Hotel was dirty and inefficient and full of white tykes... rowdy groups were always drinking and hanging about in the entrance hall.[1]

Indeed, at the time there were many remittance men in Kenya, undesirables sent abroad by their families in exchange for a quarterly allowance which apparently they regularly overspent on drink. They had a reputation for treating the Africans 'as something worse than animals' according to one set of memoirs – with aggressive use of the *kiboko* – the rhinoceros whip.[2] (Later the government seriously clamped down on the use of the *kiboko* so that by the end of the 1930s it had disappeared.)

Nairobi was poorly viewed by others too. Daphne Moore, the wife of the Chief Secretary,[3] newly arrived in 1929 from Nigeria where perhaps things were done better, described the town as a cross between a dust heap and Lyons' Corner House. The atmosphere, she wrote, 'is one of intrigue, suspicion, dishonesty and unkind criticism... the Service is underpaid, disgracefully housed and given no encouragement to do honest work'.

Yet a picture of Government Road taken in 1928 shows a wide clean street, flanked by colonnaded shops, with names like 'The Emporium' and the BEA Music store, advertising itself as the 'sole agents for East African pianos (Barnes and Mullins, Chappell Collard and Collard), and organs (Bell, Estey and Malcolm) and with strings for all instruments and new music arriving by each mail'. At the end of the road stands the fine building of the Khoja Mosque, an Ismaili Prayer Hall which had been built by public subscription and opened by the Governor Sir Edward Northey in 1920. The picture shows rows of neatly parked box body cars; the box body was ubiquitous throughout the Kenya of this period. Huxley provides a wonderful description of the box body:

> To make a box body, you got hold of a second hand chassis and built the superstructure out of planks, roofing felt and wire netting. Canvas curtains at the side were rolled up to be secured by straps, and lowered when it rained. Wire netting was often stretched across the back to keep in dogs, luggage, guns and African passengers who crowded into every vehicle that moved.[4]

Every family had stories about their box bodies – the 'flying bedstead' according to friends.

The New Stanley Hotel may have been full of undesirables, but it did claim to be the only hotel north of Johannesburg with a bathroom attached to every bedroom. There were several good hotels. In 1927 Abraham Block bought and transformed the Norfolk Hotel and he 'went on to build the finest hotel chain in the country'. In 1930 Captain Grogan, early pioneer, and early leader of the Colonists' Association with Lord Delamere, had just completed the fine new Torr's Hotel, with the only sprung dance floor in Africa. Grogan himself was one of the more outlandish characters in early Kenya days. He had arrived in 1903 but his reputation preceded him. In 1899 he walked from the Cape to Cairo to win the hand of his bride by proving to her father that although he had no money he had other resources. The walk was indeed gruelling, taking him through territories of ferocious crocodiles and cannibals, and he did indeed have courage. But although he was, according to Elspeth Huxley, 'a charmer, a cynic, a swashbuckler, a buccaneer born out of his time; generous with money; bold in his commercial ventures', she adds that 'when it came to fortune seeking, his scruples weighed (she should guess), about as much as a grain of sand'. He was a good-looking man with flamboyance and wit, but his own assessment of himself is shown in his response to a toast raised to him in 1911[5] as 'the baddest and the boldest of a bad, bold gang'. He enjoyed holding court in Torr's – with silver tongued oratory he set out plans for a Kenya ruled by the white settlers.

Grogan was determined to make Torr's the 'Carlton of East Africa'.[6] The hotel put on many amusements and events and it was in fact the first hotel in Africa to offer dancing every night; he imported a whole five-man band from Soho. Its leader was a Jewish saxophonist named Sid Zeigler and on drums there was the young Micky Migdoll. Grogan paid well for the best; the young Micky, born in Wardour Street and then only 16, received £60 per month and 'all found'; apparently this was twice as much as Legislative Council (Legco) officials received! Migdoll stayed on in Nairobi until his death in 2009 – although when war broke out he was sent to Madagascar and hated it – 'dreadful place, I nearly died there' – but as his chronicler notes, he did not, and made it back to Kenya. He became leader then of his own band, still at Torr's and later set up a men's clothing business 'Esquire' which went on to become very well recognised and apparently made suits for Sinatra, Crosby and the like. He married a beautiful Scottish woman who, after the war, had come to Kenya as a secretary for Shell. She became a highly successful trainer of racehorses, and for many years wrote a racing column in the *East African Standard* as 'The Squirrel'. Gloriously flamboyant, she is remembered as wearing enormous hats, and raising money for Barnados Homes outside the *dukas* of Nairobi. I met her in 2010, an almost touchable link to my father, and she was fiercely rejecting the idea that age should limit her – 'why do they keep saying "at my age" – there is nothing wrong with me?' She is still elegant and beautiful and still cares for her horses, but no longer walks them along the road from the stables to the racecourse.

Music and theatre were strong in Nairobi in the 1930s, and after his three years in piano-less Meru, Owen threw himself into the vibrant cultural scene. In October 1933 the Torr's Hotel Orchestra (Sid Zeigler on bassoon, L. S. Whiteson on violin, and Micky Migdoll on tympani) joined the Nairobi and District Philharmonic Society Orchestra for a promenade concert under the baton of Billy Isherwood. Owen Sunman sang the prologue to Leoncavallo's *I Pagliacci* (the critic in the *East African Standard* noted that 'he… gave a good rendering. But his strong voice would have sounded to better advantage had he remained on the concert floor instead of going to the upper balcony from where the ceiling threw his voice back at him') and the programme included Schubert's Unfinished Symphony, Mendelssohn's Violin Concerto in E minor and Gershwin's *Rhapsody in Blue* as a piano diet.

Beside Torr's was the Theatre Royal, one of Nairobi's two theatres; the other the Empire Theatre. Between them they presented an energetic and ambitious set of programmes. In 1933, there were at least four theatrical shows, and Owen, revelling in the musical vitality of Nairobi participated in all of them. First a performance for two nights of *Hiawatha* (Owen in the chorus); then a production of *The Wandering Jew* ('the most ambitious undertaking in the story of the stage in Kenya, and those who had the wisdom to see it were frankly astonished that so young an African colony could present it with such a great measure of general acceptance' as the aforementioned critic gushed) in which Owen played Raymond of Toulouse. In Gordon Daviot's play *Richard of Bordeaux*, which opened in Nairobi while it was still playing in London – 'a wonderful play, wonderfully played' – Owen took the role of Robert de Vere and 'contributed delightfully to the lighter side of things in the banquet scene and played most effectively in his final tragic scene with Richard'; and *Hamlet*, a Shakespeare's birthday tribute on 23 April, saw Owen as an officer while leading the orchestra from the viol basso. The production was acclaimed as 'a triumph' and 'A Kenya achievement'. Owen gets a mention here too – 'the only two friends of Hamlet in the whole brilliant court, the stout Horatio… and the leader of the band of players, Mr O. Sunman, bring a touch of stability and sincerity to a scene of intrigue and suspicion'. And, in March of the same year, a performance of *The Mikado* was staged at the Empire in Nairobi – Owen Sunman here was 'effective in his small part of Pish Tush' – and again the play was deemed 'A fine production'.

In fact, this was the third production of *The Mikado* in Nairobi – the first had been in 1913, and the second in 1927. Several of the cast from the 1927 production reprised their roles in 1933, but only Theo Blunt (an Australian soldier who had arrived in Kenya sometime after the Boer War) was in all three, in the role of Koko the Lord High Executioner. He went on to play the role again in 1950.

There may have been more productions, but I only have records of these as Owen only kept the programmes and reviews of productions he performed in – all autographed by the other members of the cast. Nairobi must have been a terrific change after the forested depths of Mount Kenya for someone for whom music was his life-blood. In fact, such was the artistic activity that one exhausted letter writer to *The Standard* complains at the mass of theatrical and musical entertainment. In 1933 he writes:

> Take for example the period commencing September 6th. It is the opening date of 'The Wandering Jew', played by amateurs; the same evening the dance championships take place at Torr's Hotel; at the Capitol is to be shown the British film 'Nine to Six', a first release in Africa; while the Empire will show 'Hell Divers', one of the most spectacular of American films.[7]

Clearly it was difficult to choose, or perhaps it was like London buses – you wait for months in a cultural desert and then five arrive all at once.

While there seems to have been a cultural flowering in the late 20s and early 30s, there had been musical and theatrical endeavour in Kenya for some years. As far back as 1912 one Marie Shorten performed at the Lyric Hall in Nairobi and 'returned world famous' to the Lyric in 1926, and the appreciation of Kenya as a musical centre meant that other well-known artists – the cellist May Mukle[8] and pianist Marie Hall for example – also visited in 1927 for the first Kenya Music Festival. It was to Kenya's advantage that it was en route to South Africa and a suitable stop-off point for artistes en route. A vaudeville troupe was formed in 1922 at the Lyric Hall, and an amateur operatic company was formed in 1926–27 and productions of *Iolanthe* and *The Gondoliers* and, of course, *The Mikado* were shown.

There was also a serious musical world in Kenya. The Nairobi Philharmonic Orchestra was set up in the mid-1920s by Billy Isherwood, as a successor to the Nairobi Orchestral Society which had been established in 1918. The society's president was John Ainsworth, and Mrs Ainsworth sang one of the solo roles in the later production of *Hiawatha*. The fact that, while Billy Isherwood was clearly the leading musical light, the timpanist from *Hiawatha* (which Isherwood conducted), E. M. Ley, had conducted the orchestra to *The Mikado* earlier in the same year, suggests that there was some depth of talent in Nairobi.

Billy Isherwood was one of the driving musical forces. He arrived in East Africa with his father in 1907 and for some time worked in his engineering business. He had various métiers: a diamond prospector, a building inspector for the council. He was later elected to the Municipal Council in 1919, but it was for music that he was most renowned. He had a fine baritone voice, and he played the organ at St Stephen's Church when he arrived. His wife Deborah had a fine soprano voice too. When the All Saints Cathedral – originally known as the Cathedral of the Highlands – was built (the foundation stone was laid in 1917), Billy Isherwood became the organist. He later moved out to Limuru[9] where he became the organist and built a large, quality choir, the basis for the Nairobi District Choral Society. He founded the Nairobi Philharmonic Society in the mid-1920s and, with his choir and local instrumentalists, conducted many works, including *Hiawatha* and the Promenade Concert that Owen sang in. Over the next 20 years, the Philharmonic Society gave performances of all the well-known oratorios and orchestral pieces. He and Owen must have been soul mates sharing a passion for music making and particularly the fine English tradition of choral and church music. Later, in 1943, while Owen was building up the choir at All Saints Cathedral, Isherwood founded the East African (later Kenya) Conservatoire of Music which, despite many vicissitudes over the years, is still going strong today – as is the choir.

In 1939 the Nairobi Music Society was set up and chaired by Jean Laird Bruce, a professional singer – she was apparently a magnetic Katisha.[10] Owen also sang in the cathedral choir, and during 1934 the cathedral staged a 'Pageant of East Africa', a dramatic venture showing the history of East Africa from the Portuguese landings in Mombasa to the present day. Interestingly, the Goan community contributed strongly to this musical pool. In the 1920s there was a well-received Goan orchestra and the Goan Music Society was also thriving in the 1930s. In the scene of the New Year's Eve Ball in the film of *Out of Africa* the music was – authentically – supplied by a Goan trio. The D'Souza bar in Nairobi was a centre for music making, and while in later years the Philharmonic and the Goan orchestras dwindled away, there was for many years a good representation of Goans on strings in the Nairobi Orchestra through to the 1970s and beyond.

As well as the larger pieces demanding choir or chorus and orchestra, there was a lot of chamber scale music making, in Nairobi and also up country. Again, A. J. R. Master who actually produced *Hiawatha* in 1933 writes that in the late 1920s, 'frequent visits from country district companies indicated that the Colony was bursting with so much artistic talent that the existing organisations were unable to provide sufficient outlet'. So on the one hand Nairobi was the playground of the notorious 'Happy Valley set', home of remittance men and other 'tykes' but also a city of some arts and culture. By the end of the 1940s it even had its own eponymously named *Art* magazine, which covered not only theatre and music, but contained articles on John Piper's work and discussed (approvingly) the placing of modern sculpture (Picasso, Moore and Hepworth) in London parks.

Sir Edward Grigg, governor from 1925 to 1931, played an active role in supporting the arts in Nairobi, and serious theatre began to flourish. The leading theatrical group was the Railway Players, set up in 1933 by members of the Railway Institute,[11] but welcoming non-railway participants. E. B. Pasmore, who produced and starred in *The Wandering Jew*, was a leading theatrical light, and was instrumental in focusing the Railway Players on serious theatre, despite scepticism among some Nairobi residents: 'An announcement that the [Theatre Royal] was to be reopened for the presentation of worthwhile plays attracted immediate support, although some people shook their heads and said Nairobi wanted only frivolous types of entertainment'. But it was a great success, and at the first production, *Outward Bound*, many people had to be turned away from performances because they had failed to book in advance, and the local papers gave the venture 'such excellent notices that future support was assured for the front-rank works listed to follow'. These included the above mentioned productions of *Hamlet* and *Richard of Bordeaux*.

Artistic endeavour provided a meeting place for the different European tribes of Kenya – the settlers, the government servants and the business community. As would be expected, some cast members were civil servants and engineers: Pat Manley, who played Laertes was chief officer in the Prisons Department; there were several individuals who were engineers in the Railways Department; and two government auditors; and the chief legal adviser to the PWD. But in the cast list of *Richard of Bordeaux*, Owen's name is next to that of Glady Delamere, second wife of Lord Delamere.

Glady Delamere, (who was also known as Gwladys), was in some sense the social queen and Tsarina of Nairobi, and had married Lord Delamere in 1928 when he was 30 years her senior. A striking woman with pale skin and dark hair she is described as 'good at putting people at their ease' and had a reputation for throwing excellent parties. Huxley writes of her that 'people reacted positively to Glady; either they liked her or they did not' and this is reflected in the range of comments about her. On the one hand, Karen Blixen in a letter to her sister was critical of Glady. During a visit while Edward Prince of Wales was in Kenya, Blixen writes:

> Lady Delamere behaved scandalously at supper, I thought. She bombarded the Prince of Wales with big pieces of bread, and one of them hit me, sitting beside him, in the eye... and (she) finished up by rushing at him, overturning his chair and rolling him around on the floor.[12]

(Note, however, that Blixen had secured a place next to the Prince; local cattiness commented that she spent a fortune on dresses for this occasion, considering that as a Danish subject she was not entitled to dine at official functions in Kenya.) Among other escapades, Glady had a brief affair with Lord Erroll, who was later murdered and there were some who suspected her of the murder. On the other hand, Daphne Moore, who had been so disappointed in Nairobi, said that 'she was charming and clever and suggests Elinor Glyn with sofas strewn with tiger skin'[13] and she had been a friend since childhood of

Joan Grigg, the governor's wife and, like her, was very involved in good works. In 1934, after Lord Delamere had died she was elected with a handsome majority (236 votes to 80) to the Municipal Council where she was an able and energetic councillor. She was later elected Mayor. Theatricals are not mentioned in these writings about her, but in *Richard of Bordeaux* 'as the housewifely Mary Bolingbroke [she] gave a quiet and restrained performance which carried considerable weight'; another reviewer commented that '[she] demonstrates her versatility by playing the part of the prudish Mary Bohum, Countess of Derby'. Clearly the reviewer was aware that 'prudish' was not an adjective which could be applied to Glady.

Some of these thespians stayed on to become leading lights within Kenyan society. Humphrey Slade who was also in *Richard of Bordeaux* played leading roles in a number of productions; he was a lawyer with Hamilton Harrison and Matthews in Nairobi, and later, at the time of Independence was unanimously elected first Speaker in the Parliament. John Rawlings, tenor soloist in *Hiawatha*, was an accountant with Fenton and Rawlins and in the 1950s became organist and choir master at the cathedral in Nairobi (where Owen was himself founder choirmaster in the 1940s). Theo Blunt, whom we have already met, many times Koko, became a successful businessman after his start in the PWD, and enjoyed his success so much that he was said to drive the 100 yards from his home to the bar at the Norfolk Hotel for a gin and tonic in his Rolls Royce. A. J. R. Master continued producing theatrical attractions in Nairobi for many years, and helped to build up the Shakespeare Festival – the 1933 *Hamlet* was the first production in this series.

Some worked in the artistic and media world. Kathleen 'Goody' Robinson who had played Gertrude in *Hamlet* had come to Kenya as a girl in 1928, and was described in her obituaries as 'an accomplished actress and respected journalist and broadcaster', as well as assistant editor of the *Sunday Post*, and leading light in the establishment of the New Theatre. John Mallett who played the title roles in both *Richard of Bordeaux* and *Hamlet* is remembered as trustee of the scheme in the 1950s to build a National Theatre (which still stands). Arthur Devas-Jones who took the part of the Ghost in *Hamlet* began life in the consular service, but later became a well-known and lively freelance journalist and drama critic, also with the *Kenya Sunday Post.* Stanley Daly, a page in *Richard of Bordeaux* and at first an insurance agent, played with Sid Zeigler – who turned up later in Kakamega during the gold rush.

Joan Martin who sang Pitti Sing in *The Mikado* was married to Jean Remi Martin. Born in Manila he arrived in Kenya in 1928 to work in the family business of Tyson Brothers Ltd – agents for Arbuthnot Latham and Co., London-based merchant bankers. The bank at that time was financing the Karen Coffee Company, through which Baroness Blixen operated her farm. In early 1931 it became evident that she could no longer carry on and Remi Martin persuaded his family in the UK to put up the finance to develop the property as a residential development. The contract was signed, but before she left Blixen was quite generously offered her house and 20 acres – to which she retorted that she would rather live on one acre in the Sahara than on 20 acres in a Nairobi suburb. But this was the beginning of the suburb of Karen, and Martin went on to found the Karen Club and build and design the golf course.

The cultural world of Nairobi brought together different sets of Kenyan society who would not necessarily see much of each other – for Kenya was a colony full of persistent tensions between settlers, the government, and officials, and the other communities and there was a complicated triangle of loyalties and priorities. The early settlers' objectives were the clearest cut. To varying degrees they saw the role of government as supporting them in their objectives of building up a 'white man's country', and very early demanded a say in the government of the colony through representation on Legco. Their attitude to officials seems to have been one of disdain and superiority. They were described by one settler as puppets to be manipulated. The aristocratic early settlers, who saw Kenya as a rich man's playground, and a sportsman's paradise, could pull strings in London and felt they could effectively influence the government representatives in Kenya. They despised efforts by London to maintain decent standards of behaviour among the officials, particularly the district officers, the backbone of the colonial administration. Once, following a scandal of sexual misconduct by a number of district commissioners, Lord Crewe, Secretary of State for the Colonies, issued a circular warning members of the Colonial Service that they would face professional ruin for having 'immoral relations with native women'. The settlers' mockery took the form of a verse sung to a well-known hymn tune:

> Pity the poor official.
> When 'ere he gets a stand
> He may not have a bibi [mistress]
> He has to use his hand.[14]

The settlers had in fact earned a reputation for depravity and decadence; there were lurid tales of a community that 'drank sundowners until sunrise, took cocaine like snuff, and swapped wives so often that no one could remember the ladies' latest surnames'.[15] The notoriety culminated in the infamous murder of Lord Erroll in the early 1930s, immortalised in the book and film *White Mischief*. The Happy Valley set (so called because they had land and estates in Njoro, by the Wanjohi River) were indeed extreme. This was the set in which Glady Delamere played a significant role and it is hard to imagine her having much contact with junior colonial servants outside the freer atmosphere that theatre and music provided. Even the clubs in Nairobi were largely (informally) segregated, with the settlers dominating the Muthaiga Club, and the government officers passing their time at the Nairobi Club.

Still, the majority of settlers were hard working, and by no means rich. By 1930, there were just over 2,000 European farms, and the average farm size was around 1,400 acres, very different from the vast holdings of Delamere and Grogan. The distribution between the land owners was very uneven – only 13 individuals owned one fifth of all the leases issued. But the mix of characters was changing; one commentator (A. E. Harrar) wrote that 'a number of settlers of a much better class (i.e. not the aristocrats of the earliest days) had established successful farms'.[16] There had by now been some successful soldier settler schemes; a Land Bank had been established and schools had been opened for European children. Most of the farmers worked extremely hard, living through risk and uncertainty: the Great Depression, when export prices had halved, had a terrible impact on many farmers; banks foreclosed and bankruptcy was not infrequent. Many farmers lived a very precarious existence.

In the light of the hardships of the farming life, the settlers envied the seemingly generous conditions under which government officials were employed. They – even the most junior like Owen and his contemporaries – arrived in Kenya to find housing and staff awaiting them; their salaries were comfortable, they had care when sick and regular home leave at the government's expense. They also had good sports facilities, and were encouraged by their employers to make use of them. Planters and settlers resented this and would do their best to resist increases in salaries for government officials.

They were also, less justifiably, jealous of the officials' competence in the local languages which gave them, it seemed, an unfair advantage in their ability to communicate with the Africans. Government officials were of course required to learn the local languages – Swahili mainly but also the languages in the various regions where they were posted – and their salaries and promotion were tied to their linguistic competence, whereas many of the settlers spoke only a kitchen Swahili, and did not try to learn. The records show that Owen had completed his preliminary exam in Swahili before he departed on leave in 1930. Perhaps Elspeth Huxley's father typifies the settler attitudes to local languages: she wrote 'although he was an excellent linguist, with fluent French, German and Spanish, some curious barrier seemed to exist between him and our infinitely simpler up-country Swahili... he did not bother to learn it'.[17] These resentments inhibited trust between settler and official; the settler never seemed to understand, writes Harrar, that the officials were there to serve and help him. But the contrast between the security of the officials and the risk and hard work experienced by many of the settlers would have been annoying to say the least.

At a policy level, the key areas of tension between government and settlers were threefold. First, settlers continued to demand more forcible access to labour – the Hut Tax had gone some way to ensuring a ready supply of labour, but there were those among them – led by Delamere and Grogan – who demanded that 'the amount of land the natives were allowed to hold should be absolutely limited' and that they must 'force the native to work' through mechanisms like the whip and Transvaal style Pass Laws. On this issue, the government upheld the rights of native labour, insisting that natives need not work unless they wished to do so, and that native chiefs under persuasion of the settlers had no right to compel labour to go out, a practice that was prevalent in the earlier years of the twentieth century. Predictably, the Colonists' Association was furious, arguing that the restrictions at least doubled the cost of labour – which may well have been the case. A second issue lay in the granting of land to others – the settlers always resisted any granting of land to the Indian population, and, in cahoots with Governor Grigg, fought the drafting of a bill, the Native Land Trust Bill, which would protect the land occupied by the native population effectively.

A third issue concerned the intertwined demand of the settlers for representation on Legco, and their determination to minimise the rights of other communities. From the early days, under the leadership of Lord Delamere, the settlers sought to secure their own privileged position (access to land, access to labour) and, conversely, to ensure that other communities, specifically the Indian, would have minimal rights and roles in the community. They demanded a formal role in the allocation of land and political representation in government, and in 1906, in response to the demands of Lord Delamere and Captain Grogan (who would succeed Delamere on his death as de facto leader of the settlers), Lord Elgin, secretary of state for the colonies in the new liberal government had conceded that there should be settler representation on a legislative council in Kenya. This was not universally welcomed, as the comment by his under secretary, Winston Churchill suggests; 'Never before in Colonial experience has a council been granted where the number of settlers is so few' – there were about 6,000.

In fact just before the unofficial members were nominated, in March 1907, a shocking incident occurred. Captain Grogan, who was by then President of the Colonists' Association, set up to defend and extend the settlers' rights, took a leading part in the public flogging of three Kikuyu servants in front of the Nairobi Court House. This was seen as a deliberate challenge to the government, and was taken so seriously in London that a Parliamentary Paper was published on it. Many officials, both in Nairobi and at the Colonial Office, urged that Legco should be withdrawn. But in fact the then commissioner, Sadler, recommended that the council should go ahead, since it seemed that the flogging was the work of only a small group of South African influenced settlers. The council did indeed go ahead – with one change, that Grogan was not nominated, although as president of the Colonists' Association he would have expected to be.[18]

But it took longer for the political status of the Indian community to be resolved. The Indian community also demanded representation on Legco – a reasonable request since at the time there were about 23,000 Indians in Kenya compared with 6,000 settlers – but of course this was highly inflammatory to the settlers. Whitehall was in fact adamant on this issue; Winston Churchill, by then secretary of state for the colonies, strongly supported Indian representation, arguing in a minute in 1907 that: 'There can be no reason for excluding this large and meritorious class. Begin early to instil good principles in the East Africa Protectorate'. Definitely a Churchillian growl. Shortly after this he visited East Africa and, far from convinced by the Colonists' Association's assertion that Kenya was a 'white man's country', he strongly supported Indians and their contribution and, in referring to 'the blacks' it was, he said, '*their* Africa'. Indeed he publicly said that there was 'ample land for all' in East Africa.

And despite the settlers' efforts, A. M. Jevanjee, the Gujarati businessman and developer who set up the statue of Queen Victoria in the Jevanjee Gardens in downtown Nairobi, where it still stands, was appointed to Legco in 1910.

During this period, Whitehall played a strong role in protecting and defining the rights and needs of both the Indian and the indigenous populations. But Whitehall and the colonial governor did not necessarily speak with one voice, and within the colonial government there was a considerable level of tension. Some governors – Charles Eliot at the beginning of the century and later Edward Grigg in the 1930s, for example – were seen as very much on the side of the settlers, and very close to them socially. There are many descriptions of this intimacy in social events both at Government House and the Muthaiga Club. This attitude was resented by those officers, such as John Ainsworth, now the first chief native commissioner, and his colleague Oscar Watkins who both saw the government's role as one of trusteeship for the African, rather than one of facilitating the life of the settlers. These tensions came out strongly in agricultural policy where government yielding to settler requirements seemed to contradict many of the goals that officers were working towards. The officers in the Department for Native Agriculture were trying to encourage African farmers to grow better cash crops as well as improving their standard of living through better husbandry but the Colonists' Association actively challenged the right of African farmers to grow anything but subsistence traditional crops. African farmers were incensed at these attempts to limit what they could grow in the Reserves and coffee, which was forbidden to Africans under the Native Grown Coffee Rule, caused the first major flash point over cash crops in a stormy row between African farmers, settlers, missionaries and officials.

In fact, the First World War had brought many profound changes in the relationship between the African population and the Europeans; during the war many Africans had fought alongside British officers, and were treated as fighting men. They were not prepared to revert to the pre-war subservient status when they returned home, and the earliest African political movements date from this period – particularly Harry Thuku's Kikuyu uprising which emerged through the Kikuyu Association and the Young Kikuyu Association, concerned with the grievances of labourers, as well as the request that proper title be issued for *githaka* holders.

Then, in the early 1920s there was a key political decision with profound implications for Africans and settlers alike: the 1923 Devonshire Declaration. The role of the Indian population in the colony was still a burning political question, ten years after the first settler representation on Legco, with the European settlers and the Indian community ranged on either side of a political divide. In 1923 a conference to resolve the issues was called in Kenya under the chairmanship of the Duke of Devonshire, secretary of state for the colonies.[19] The resulting White Paper surprised the protagonists and formally drew attention to the importance of the future of an estimated 2.5 million Africans in a colony where 10,000 whites and 23,000 Asians hustled for political power. The Devonshire Declaration pronounced that 'Primarily Kenya is an African territory. The interests of the African native must be paramount'. Thus was the government's trusteeship for the African population enshrined. This formula postponed resolution of the political role of the Indian population, but instead established the paramountcy of the native population. In 1924 the missionary Dr John Arthur[20] was appointed to represent African interests on Legco under the provision of the White Paper. Obviously this was a long way from proper political representation, but it did, for the first time, show a formal recognition of the rights of the indigenous population. Naturally, the Colonists' Association was horrified.

The war also brought a new type of settler. There were several initiatives to encourage ex-servicemen to settle and take up farming – known as the soldier settler schemes – the export of 'Land Fit for Heroes' perhaps. By 1920, some 400 farmers had received land under the Discharged Soldiers Settlement Ordnance during 1919 and 1920. Most had no previous experience of pioneering work or of farming.

Some settlements were successful but there were some disasters and one of the most extraordinary of these was BEADOC – British East African Disabled Officers' Colony. Some 50 officers, war disabled in various ways, were selected (through an interview process in London) to come to Kenya in 1919 and grow flax. The selected servicemen were provided with an initial endowment of 160 acres each for a financial contribution of £600, to be farmed cooperatively. Land was allocated near Kericho, and an initial contingent of 13 soldiers arrived with their wives in 1919. Unfortunately, in 1920, just as the BEADOC was about to harvest its first crop of flax (the economic basis of the venture – on 800 acres rather than the initially planned 5,000) the price of flax dropped from £590 to £80 per ton, far below the cost of production. No further orders from the aircraft industry and the war machine meant that the market simply disappeared. BEADOC collapsed, and only four families stayed on to farm in Kenya.[21] But one of the farmers, Bill Lyne-Watt, who had been injured when on a training flight with the RAF, stayed on and joined the Department of Agriculture, was a forceful pioneer at the time and became a colleague and friend of Owen's.[22]

At this time, when not starring in local productions, and enjoying his bachelor life, Owen could be found at the department headquarters working on the first long-term strategy for the sector; the department undertook a systematic survey of resources, and was determined to transform the land and the farms in the native areas. A census of land use in the colony showed that the area alienated for white settlement was just over 6 million acres, of which just over half, 3.2 million acres (52 per cent), was under cultivation. At the time there were 1,182 European farm owners or managers. The census also shows that the areas for Native Reserves (excluding the pastoralists) totalled 33.6 million acres[23] of land designated for Africans, where no land was to be alienated for white settlers, and the traditional patterns of life could be sustained. The defining of reserves was meant to 'limit absolutely' the areas of land that could be alienated for the whites, and to secure traditional lands for

African farmers. This was the basis upon which the department's strategy for native farmers would be built.

Later commentators have argued that there was very little clear development policy for agriculture in the Reserves until the later 1940s, after the Second World War, and indeed, the Swynnerton Plan which was produced then was an important document. However, my reading of the department's commentary is that there was a real will for agricultural development, although it was constrained by a combination of shortages of resources, and institutional bias towards the settlers' agriculture. To some extent this was inevitable since the aim of settlement was to produce goods to transport by rail and for export and the settler agriculture was much more productive and efficient than the traditional agriculture. Over the coming 20 years, policies for native agriculture matured; in 1934 the department was still evolving in terms of activities and policies – and the technical support was thinly spread. Writings from agricultural officers during this period show how thinking developed and agricultural policy became more coherent as this cohort of young AOs matured in terms of knowledge and understanding.

But despite declarations from London of native paramountcy, mistrust and suspicion continued, and the occupation of the Highlands by immigrant settlers did not give the Africans any sense of faith in the moves to improve agriculture in the Reserves. The official writings of the time highlight the ambiguity of the colonial mission. Holm writes: 'By coordinated efforts between the Department and the district officers it is anticipated that much good will accrue, *not only to the native* but to the colony'. This phrase alone contains the confusion of agricultural development at that time: the good of the native? Or the good of the colony? Could both be achieved? This ambivalence comes out later in Holm's 1923 report. He notes that increasing demand for native labour on European holdings 'is one contribution to the prosperity of our native population and one of great value to the labourer'. But at the same time he says:

> natives, if encouraged to work and produce in their reserves will learn much, particularly the value and dignity of labour.

> As they prosper their dietary [sic] should become more varied and their lives lengthened. Improvement in the quality and physique of future generations should be manifest.[24]

There is a clear tension here between the need for labour on the settler farms and the desire to develop native agriculture in the Reserves, and it sets out clearly the context within which Owen and his colleagues would find themselves.

But it was hard work and, like the settler farming, it involved experimentation and trial and error. As Holm had commented in his annual report in 1925:

> only those closely associated with the day to day administration and activities of the Department of Agriculture, serving a country so young in its stage of development and about which so little is known, can fully realise how inadequate is the information and knowledge possessed of its resources, the lines along which development should proceed and the best methods to be adopted.[25]

In 1933 Alex Holm, the pioneer director of agriculture, had to retire from his post due to ill health. Owen sent copies of the various tributes to him in the local press back to Wye College, where they are reported in the *Agricola* (alumni) magazine (Holm had been a founder member of the Agricola Society when he graduated from Wye in 1897). The tributes describe a man of serious commitment to the agriculture of Kenya across the board, in both the White Highlands and the Native Reserves. In his role as director for agriculture he held responsibilities in government as a member of both Legco and the Executive Committee (Exco). He also set up the Royal Agricultural and Horticultural Society and served as its president for 14 years. In the leader of the *East African Standard* in December 1932, the tributes to him noted specifically that 'soon after his arrival in Kenya, the Director realised the big potential value of native agriculture and although developments in this direction are necessarily slow, a sound foundation has been laid'. One of his achievements over the 14 years of his service, was to treble the budget allocated to native agriculture until it accounted for about one fifth of the total.

When Owen left on home leave in March 1934 he was certainly expecting to continue working in Nairobi on his return. But on his return things were quite different. For one thing, he was immediately posted to Western Kenya, and for another, he brought back the girl next door as his bride. This too was unexpected, as she was already engaged to a nice young man called Norman. But their story – Owen and Dorothy's – is one of a whirlwind. Against this dashing, glamorous, young colonial officer, poor Norman did not stand a chance.

The Sunman family, William and Alice, both teachers, and their daughter Josephine (Jossie), lived in Minster Road, Bromley, Kent. In the early 1920s the Adams family moved in next door and the two families became friends. The families had a lot in common, both hardworking non-conformist (the Sunmans were Congregationalist, the Adamses Methodist) families with a sprinkling of clergy and missionaries among their forebears. The Adams family were wealthier, however; Charles, Dorothy's father, was a prosperous businessman in the City of London where he ran the eponymously named 'City Safe Deposit'. Dorothy had a younger sister, Joan, who was a quiet, shy child. Life ran along orderly lines: Charles took the train every day to the office in Throgmorton Street, and took a (small) gin and water on his return. Theirs was an austere non-conformism. Once Dorothy said she had seen her Uncle George going into a public house and it caused outrage in the family, she was never to speak of it again. In fact, George was disinherited by his father for drinking, and only total abstention would restore to him his heritage. Such an attitude was not unusual in non-conformist circles, and indeed many people 'signed the pledge of temperance' – a pledge of teetotalism that had been part of the non-conformist movement since the mid nineteenth century. Jossie, Owen's sister, was one of these and she signed the pledge during the 1920s.

Dorothy was now 26, working as a secretary in the City of London. She was not an academic girl but full of practical good sense. She was a leading Girl Guide in her community in Beckenham and when she had gained her school certificate she took a secretarial course at the City of London College in Moorfields, staying in lodgings near Old Street. Her teacher was, by her account, something of a dragon who would rap the girls' knuckles with a ruler if their touch typing was inaccurate or not up to speed. The course covered shorthand, typing and book keeping – an area she excelled in and she was always proud of her ability to cast a sum across a sheet of accounting paper faster than her colleagues, and always accurately. She took a job in a law company in Throgmorton Avenue – close to her father's city safe deposit company, and it was while she was working here she met her fiancé Norman. He was a pleasant, respectable young man; an accountant. Her life was ordinary and suburban. By mid-1934 they were set on a safe and comfortable path, with a wedding planned for early 1935.

According to Dorothy's version of events, she had regarded Owen as a rather bumptious boy next door, with an overabundance of self-confidence; an acceptable neighbour but certainly nothing special. This is certainly consistent with comments from his school which reported, ten years earlier when he was 18, 'his work is good but not good enough and certainly not as good as he thinks it is; I regret to feel that the phase of innocent self conceit through which many boys pass has lasted too long in his case'. It was a little earlier, when Owen was about 16 and Dorothy 14, that her family moved to Bromley Road, Beckenham, and while the parents kept in close touch, the children did not – although they had all gone to see him off in 1928. On his first home leave in 1930, Owen regaled his family and their friends and neighbours with tales of travel and adventure – safaris in Meru and Embu, fantastic landscapes, elephant and buffalo, bridge, tennis and golf.

Apart from the stories Owen brought back with him and the gossip between the two families, Dorothy would have been well aware of empire. In the early 1930s, empire and colonial service held a central place in the culture and life of the British; touched with a certain glamour, a combination of exoticism and philanthropy. Through exhibitions, magazines, documentaries, contacts with Women's Institutes and literary societies, the British Empire was portrayed – and understood – as a 'single family of diverse yet united peoples'.[26] There was a direct appeal to women – the Wembley Exhibition had a Women's Section and a Women's Week, the later Glasgow Exhibition (1938) had a separate pavilion illustrating the work, products, and handicrafts of women throughout the Empire. The BBC was rich in stories of peoples and places. Her Methodist background gave Dorothy a commitment to community service, a missionary enthusiasm to improve the lot of the poorer members of the Empire. Whether or not she had any yearning to be involved in this life, when Owen returned home on his second leave in 1934, full of more stories this time including the high life of Nairobi, the music and theatre and dances, Dorothy – to her surprise – was bewitched.

So was he. There are few diaries of his life but a small notebook of his courtship of Dorothy describes a true love and passion. The courtship was swift. Owen arrived back in the UK on 30 March, and their engagement was announced six weeks later, on 8 May. In July they went to the Cricket Week Dance at Wye – and he took her to a performance of *Gotterdammerung* at the Royal Opera House. They sat for five long hours on wooden benches in the balcony; this was not repeated, Dorothy's tastes were less highbrow than that. They also spent a week together with Owen's family in Cornwall, where Alice taught my mother to play bridge (she was fierce). Whether there was any more practical training is not clear, but bridge was an essential.

It is hard to imagine how all the preparations could be made in such a short time. A wedding can take months to prepare: guest list to draw up; invitations to send; wedding outfits to be designed and bought. But on top of this was the preparation for life in Kenya. The Colonial Office was nothing if not practical and again lists were available to help the new young bride prepare herself for Africa. One such list goes to great detail on garments and under garments: dressing gown and slippers, not too glamorous; four to six nightgowns or pyjamas; four to six cotton or linen or washing silk day dresses and a smart day dress for official occasions and lunch parties. Three evening dresses, and games clothes, and so on. These would be common sense, but in addition brides were advised to take knitting patterns, a couple of plain adaptable patterns for day dress and evening dress; knitting needles and a compendium of underwear patterns. A nice useless piece of needlework, gros point or a luncheon set or tea cloth – to fall back on if you were ever idle. There was little to buy in Kenya despite the sophistication of Nairobi and clothes were to be handmade at home or by a dress maker, as were curtains, cushion covers, and the niceties of comfortable living.

They were married on 5 September 1934 at the Methodist Church in Beckenham; Dorothy wore a white georgette draped dress with a romantic floating bouquet of roses, gypsophilla and blue scabious; on her head a small seed pearl tiara, and round her neck a white pearl necklace. (I wore that for my weddings too.) Her wedding ring – which I wear now – was a thin band of platinum, and her engagement ring was a small cluster of diamonds. The wedding pictures show the family group: Charles and Ethel, Dorothy's parents and her younger sister Joan, and William and Alice Sunman and their daughter Jossie; the girls were bridesmaids, together with a friend of Dorothy's, Jean Thorogood. They wore dresses in pale blue with wide brimmed, floppy hats; a smiling happy wedding group.

Dorothy and Owen evidently were happy and look in total harmony together. But it must have been a time of mixed feelings for both families; Owen's had become accustomed to his absence and the regular exchange of letters; but close as the families were it must have been bittersweet for the Adams group, who until a few months earlier had been expecting that their older daughter would be settling down happily nearby in south London or in Kent with the respectable and reliable Norman; happy, they would have presumed, and ready to start a family and maintain rituals of regular Sunday lunches together, sharing the life of the Methodist church. Now she was uprooted and sailing off – just three weeks after the wedding – to Africa, a very different land, a very different life. Joan adored her older sister and, a nervous child, would feel the loss intensely.

The wedding photos show a close group; Owen's best man was Eric (much later to be Lord) Ashby, and his younger brother Maurice was an usher. Owen had got to know the Ashby family through the local music society, and they remained close friends for the rest of their lives. In fact, many years later, after my father had become a vicar, when I married, he officiated and Eric Ashby gave me away, *in loco patris*.

The wedding presents were laid out on tables: a dinner service and silver tea service; a mantel piece clock with Westminster chimes; a writing bureau; a full length mirror; a sewing machine in a wooden case; a pedestal desk; a travelling trunk with DES – Dorothy's new initials – painted on it; and, curiously, a Tilley lamp. Owen's parents gave them a piano. All these were immediately packed up for transport off to Kenya. There was also a Visitors' Book – thanks to which I have been able to trace where they lived in Kenya, and who stayed with them where and when. It also lists all the people who came to the wedding – about 80 of them – starting with Charles and Ethel Adams, followed by William and Alice Sunman. He wrote his address as 14 Minster Road, but Alice, fiercely Cornish always, gave her address as 'Pedna Vounder', Bromley.

Cornwall was always important, and Dorothy and Owen spent their honeymoon at Mevagissey; then, only three weeks later they were off on the SS *Madura* and arrived in Mombasa on 18 October. Photographs show Dorothy and Owen on the ship, both in white suits, he in a pith helmet while she wears an elegant cream cotton hat with a wide brim.

As an engagement present Owen had bought her a beautiful long black evening cape in silk velvet, for the cool but sociable Nairobi evenings. But sadly, this was not to be of much use; he was immediately posted to Kakamega in Western Nyanza, a far cry from the glamour of theatre, concerts and dances in Nairobi, so the cape was eventually made into a long sleeved evening jacket (I still have and wear that too).

By the time that Dorothy made the train trip from Mombasa in 1934 to Nairobi there was a restaurant car on the train, so no more hasty meals snatched at unlovely station stops; although there are no records of the quality of food on the railway at that time. Dorothy never cared for the heat and found Mombasa trying, and the journey hot and dusty. They stayed in Nairobi for a day or two before they set off to Kakamega, a night or two at the Norfolk Hotel with perhaps a visit to the Nairobi Club for drinks with Owen's friends. The climate in Nairobi in October is like a good English summer – temperatures in the 60s and lower 70s, but cooling at night still – fires were needed in the evenings.

The journey continued by train to Kisumu through mesmerising landscape. Climbing out of Nairobi towards the Rift Valley, the train crept up through Limuru to a height of 7,500 feet, a climb of 2,000 feet until, at the edge of the valley, the land dropped almost vertically by 1,500 feet, a drop of one-third of a mile. On the other side of the great brown and green valley, 100 miles away the blue outline of the Mau mountain range was drawn against the vast African sky. Animals roamed in the valley, giraffe and zebra lined the railway line. Then abruptly the train plunged, weaving back and forth down the great escarpment until it reached the valley floor and the blue waters of Lake Naivasha. This was a suitable place for a break, a leg stretch; and perhaps a meal. Then the train followed the valley floor until it reached Nakuru when it had to begin to climb again, over the Mau, even higher now. This part of the journey showed off the beautiful openness of Kenya – fertile land, sweeping pastures, endless vistas. They stayed overnight at Nakuru and resumed their journey the following day as the train moved into deep green forest, tangled with creepers and dark mystery. And the train descended again, down through the forest to the marshy wooded lands and Lake Victoria and more steamy heat.

Whether Dorothy saw and appreciated this landscape is undocumented. She never much liked heat but she would have found the wildlife and landscape extraordinary – perhaps even better than she had imagined from Owen's clear, vivid and loving descriptions, descriptions designed to entice his love to join him in this great adventure.

Photographs of their arrival in Kisumu show Owen's reunion with his car, a box body on a Vauxhall chassis; and his staff, Nyagga the cook and Mohammed the houseboy; and his beloved dog, Boris. Together they set off by road from Kisumu, to their first home together, on the *boma* in Kakamega. There is a picture of their house, a typical government two bedroom bungalow with corrugated iron roof overhanging a small porch and verandah, covered in boxes and crates as they arrive to move into their first home together. Dorothy sits on her trunk holding Boris, and Nyagga and Mohammed stand by smiling. They sent photographs home to their family: the table laid out for their first breakfast, showing a silver fruit bowl, a wedding present, as well as the essential bottles of Heinz tomato ketchup, Tabasco and Worcestershire sauce; their bedroom, quite small with twin beds and eiderdowns and thin cotton curtains that let too much daylight through and fine wooden floors – Dorothy has a small dressing table and a mirror and is doing her hair. A new life indeed.

4

WESTERN KENYA

Nyanza Province 1934–39

> 'Stated baldly, what we are doing, like the colonial service everywhere else in the world, is to accomplish a social revolution: to try to help the transition from the iron age of 1900 to standards of life and social behaviour which are set by the immigrant European community, and to effect this without imposing intolerable strains in the social fabric of the tribe with resulting violence and sorrow. We cannot claim to have achieved this in neighbouring Kikuyu. It is a great tribute to the officers here in the past, and to the good nature of the vast majority of the tribes who believe in what we are trying to do and trust us because they think it is right.'
>
> John Howard, agricultural officer, 1939

In the early 1930s, the Great Depression exerted a harsh grip on Kenya. Locusts, drought and inevitable famine, on the one hand, and the Great Depression, on the other, had conspired to create considerable economic pain. As one Kenya historian told me:

> By 1930 the price of coffee had fallen to one tenth of its earlier value;[1] as coffee takes five years to mature, so about a third of coffee farmers went broke, and bare footed white women were selling jars of home-made jam in the street. This was following a five year drought and three consecutive years of locusts.[2]

Prices of cereals had also crashed and farming looked a very insecure business. World prices for all agricultural products were depressed, and credits were made available to 'land efficient farmers whose prospects were good but whose financial position was, owing to these conditions, becoming precarious'.[3] But life for farmers is never easy, and while heavy rains in late 1929 brought food shortages to an end in many of the Native Reserves (Embu, Meru, Kitui), they also meant that it was very difficult to export surplus crops as Owen had noted; roads were impassable, and the damp maize became so infested with weevils that no one would buy it. The fall in agricultural prices affected the Native Reserves too, so that while the overall quantity of wattle and maize produced and exported increased, the value of export sales fell by a quarter. The government set in train a number of instruments to help the sector such as an Agricultural Advances Ordnance which introduced credits to 'steady a very panicky situation; and proposed the introduction of a Land and Agricultural Bank with the idea that loans could be extended to native as well as non-native farmers'.[4] But there was considerable hardship.

So when gold was discovered in Kakamega, it offered a lifeline for the country and proved a magnet for investors and adventurers. Gold had been found earlier in Kenya, near Kisii, but it was not a substantial field. However, one of the earliest prospectors there, a tall and bony American named Louis Johnson, had been a store keeper in the Klondyke before he came to Kenya, and there he had learned to use prospecting pans. There was a rumour that he had left Klondyke in a hurry as a result of a gambling episode. He found his way to Kenya via the Spanish–American war of 1898, arriving with his wife Fanny in 1911. No one quite knew about his past – he was a taciturn and silent man, whose most articulate communication was a 'clearing of the back of his nose, followed by a hefty spit, followed by the one word "Jesus"'. He and Fanny arrived in Londiani on mule back and built a stone house and farmed for 20 years in Turi, until bankrupted first by the collapse of the sisal market and then by the Depression and the collapse of maize prices. He and Fanny, a strong wiry woman with wispy hair (but reputedly an excellent hostess), set out gold prospecting in northern Tanganyika and then in Eldoret, with Gerald Smallwood, a farming neighbour and Jack Collins, a partner in the Arnold Mbarara Mines in Uganda, and funds of £100. The story goes that, after the expedition to Kisii, where they had found little, they set off for home exhausted and penniless. They camped for their last night near Kakamega by the Yala River. Before they left, they panned a tributary of the river without much expectation of success, and at last found gold. Johnson managed to borrow some more money and the four set up another syndicate with a number of other farmers, and obtained an exclusive prospecting license over a considerable area.

This was the beginning of a gold rush; once Johnson had obtained a concession and pegged out a number of claims, farmers and bankers alike rushed to Kakamega. Olga Watkins, wife of Oscar, now deputy chief native commissioner, and herself a farmer, gives a savour of the atmosphere in the early days of the rush in a letter to her mother in 1930:

> lately we have been inundated with gold miners and those concerned therein. Most interesting.... First came a man who said he had invested a tidy bit of capital (£50!) and hoped thereon to pay his £25 license and live for four or five months until his ship came home.... And then came the family who said they had been terribly hard-up and had scraped together everything they had to sink a shaft in the reef. It turned out they had scraped £50,000! How comparative everything in this world is.... Everyone who has gone broke locally seems to have borrowed or stolen enough to get to Kakamega and start, so there are a good many folk there. Quite a lot of people have been so broke that they have been living on buck and maize meal and have had no papers or anything for years. This is an adventure for men who have been cooped up with hard-up families for years with no break and no pocket money. And now they are off on their own with an open air life and the hope of a gamble.[5]

Prospectors erected posts to stake their claims, and set up washing boxes to collect gold as it sank to the bottom of the river. Shafts were dug, and by 1932 there were some 200 miners, living with their families in huts made of Kavirondo matting – reeds tied together with string. In the early days the hotels and banks too were built of this flimsy material. The finds created work for desperate farmers and encouraged traders, shop keepers and hoteliers to establish themselves in Kakamega.

Three years later there were over 600 European miners, with about 15,000 Africans working on the surface and in the pits. Africans worked mainly as labourers in the mines, but a significant number worked as drivers, blacksmiths, carpenters and masons. They were mainly fairly unskilled but Rosterman's, one of the larger companies, employed an African machinist working on lathes and shaping and planing machines who could, from a machine drawing, make finely engineered parts accurate to one thousandth of an inch.[6]

This was the peak for the small miner, who made money working the surface and then sold out happily as the big boys moved in with injections of capital: Tanganyika Concessions Ltd who sought exclusive concessions on 9,500 acres in Kavirondo and developed major holdings in Katanga in later years,[7] and Bewick, Moering & Company, a huge London based mine management company with interests from Cornish kaolin to Chinese coal mines. By 1934, gold accounted for 4 per cent of Kenya's exports, and by 1941 this had increased to 15 per cent (£696,000). A few of the prospectors did quite well, finding payable reefs or winning alluvial gold, but by the end of 1935 most had packed up and returned to their farms, 'sadder but wiser'.[8] Not Johnson however; he sold out to Tanganyika Concessions, and when he visited England in the early 1930s, Fanny wore a necklace of gold nuggets.[9]

The gold rush of Kakamega gave another cause for mistrust between Africans and the government, yet another case of broken promises over land. After many years of effort, the Native Land Trust Bill had finally been passed in 1930. It was first mooted in 1926, under the then colonial secretary L. S.

Amery, as a bill to legislate for full protection of African land, following the earlier efforts of Ainsworth, who had first established the Native Reserves when he was commissioner for native affairs in the early 1920s. But its passage was impeded by Governor Edward Grigg's well-known bias in favour of settlers which led him to try to frame the ordnance in such a way that safeguards for Africans would be weak. Grigg's bill met intense opposition from within the administration and from others: from Oscar Watkins, Ainsworth's deputy, who believed with an unwavering conviction in the rights of Africans to development; from Canon Harry Leakey, who had been appointed to Legco to represent African interests; from Archdeacon Owen, missionary activist for African rights; and from other senior commissioners all of whom realised that, as drafted, the bill did not have sufficient safeguards for the protection of African lands and interests. So while the energetic town provided markets for local produce, the government's failure to protect properly African interests contributed to on-going bitterness.

The bill stipulated that no European could occupy land that was currently occupied by Africans and was designed to give full legal protection of African land. Importantly, it was stipulated that 'the reserves are for the use and the benefit of the native tribes of the Colony *for ever*', and when the bill was finally passed three key points had been added. First, were any land to be taken from a tribe, even for public services, an equal amount of land must be given to them; second, any leases granted on the tribes' land could only be for 33 years; and third, any exceptions had to be made by the secretary of state for the colonies, in Whitehall, rather than by the governor.

But when the time came and gold became a realistic option, under Grigg's governorship the safeguards, perhaps predictably, were not upheld. The Africans on the affected areas were not offered land but were offered cash in temporary exchange. Officials apparently thought the mining period would be brief and changed the law abruptly in Legco in 1931, thinking it expedient and of little importance. But as the chief native commissioner Wade observed it was inevitable that the natives' feelings would be hurt, and that the amendment [which overruled the ordnance] would 'violate some of their cherished and possibly sacred traditions'. Oscar Watkins, not surprisingly, was furious yet again. The then DC, Captain F. D. Hislop, reported in 1932 that:

> It is idle to pretend that [gold mining] is regarded by the natives with nothing but the gravest suspicions and mistrust.... Any advantage which they obtain, such as markets for their produce and opportunities of obtaining wages all pale into insignificance beside the ever present fear of losing their land, and in addition the inconvenience and disturbance to which they are frequently subjected by the mining operations.[10]

The insecurity was never offset by cash compensation[11] and indeed it is doubtful whether the cash compensation was adequate in any sense, leaving aside the question of whether it could make up for the actual loss of land – land itself always at the heart of African identity and the lasting memories of disturbance and broken promises would bedevil plans for land reform well into the future.[12]

In fact, many administrators and officials were disturbed by Governor Grigg's close relations with the more vociferous settlers, a position that was by no means shared by everyone in the administration. He was unpopular in quite a few quarters of Kenyan society, and indeed with the Colonial Office in London on account of his extravagance as well as his neglect of African interests. Even the settlers criticised his profligacy. At a time when Kenya was struggling to make good 'the deprivations of war' he spent £80,000 on a palatial Government House designed by Sir Hubert Baker, when 'most senior officials were living in 20 year old corrugated iron shacks'. By comparison, the net expenditure of the Agricultural Department in 1927 was scarcely more, at £83,000.[13] Elizabeth Watkins writes of him thus: 'From the moment he [Grigg] arrived in Kenya, Grigg's masters in the Colonial Office recorded, in minute after minute, his extravagances, his lack of tact, his inability to obtain the best of his subordinates or the cooperation of his fellow governors'. His legacy, according to her biting assessment, was one of 'total destruction of African trust in the British administration' as a result of his 'neglect of African interests and aspirations over his five years of tenure, his reluctance to make firm legislation about the land and his willingness to encourage the settlers' wilder claims'.

But Dorothy would not have been aware of this – she would have been looking at the setting and amenities of this very new town, thankful that the government offices had now been moved to Kakamega from Mumias, an unhealthy malaria infested backwater and former slave market which had been a real hardship posting.[14] Kakamega is in a rather lovely location, situated on high ground with views of the Nandi Escarpment to the east and Mount Elgon to the northwest.

With the gold boom, the town blossomed – metaphorically and literally. Near the DC's house land was cleared and grassed and planted with eucalypts. Government buildings were improved, new thatched roofs added and the green became known as Piccadilly Circus. There were new hotels with names telling of the change in fortune – the Eldorado ('delightfully run' according to one traveller) and the Golden Hope. Gardens were planted, Barclays Bank set up a branch, the road ran through town to 'Golder's Green' and 'Hampstead Heath' and social life boomed. There was jazz at the Golden Hope dance salon with Stan Daly, Len Young, and Ron Stephens, all of whom had played with Sid Zeigler's band in Nairobi. (Stan continued to keep his day job – he was an insurance agent.) There were regular sundowner parties from six to nine in the evenings.

Less raffish than Nairobi, this was apparently a relatively sedate gold rush town; on Sundays the Golden Hope played host to church services (this dual role was quite widespread in the colony; before churches were built the club – because there always was one – would often host services and other ceremonies – although it is said that some complained of the lingering smell of booze on a Sunday morning after the Saturday night before). And, thanks to gold, Kakamega was a town with a good range

of facilities and social life by the time that Owen and Dorothy arrived towards the end of 1934. It was of course very different from Dorothy's suburban south London home. As her father-in-law wrote in an early letter, 'It must all seem very wonderful to Dorothy after her experience of the city and suburban life here'... and I believe that is how she found it.

Their house was a standard two bedroom bungalow with corrugated iron roof, and a small verandah. The sitting room contained a fireplace – not imagined from London but essential even so near the equator, and they furnished it with their wedding presents and pictures from home. One bedroom would be a guest room – an essential of the colonial life where hospitality was freely given and necessary given the distances and condition of the roads. The house had electricity, but no running water and a long drop toilet in a separate hut – often a favourite haunt of snakes. But it sat in a sizeable garden filled with exotic, perfumed flowers and trees: frangipani with its thick waxy petals and rich perfume; hibiscus; zinnias; agapanthus; and jacaranda with its beautiful lilac blue blossoms. Fruits alien to south London in the 1930s grew in the garden – paw paw and mango to be picked for breakfast, and avocado pears for lunch or dinner, served with Heinz salad cream and Worcestershire sauce. Kites wheeled and swooped; a grey parrot might be spotted in a tree. Weaver birds argued and twittered, hanging intricate baskets with woven grasses for nests. Dorothy immediately loved the fecundity and vivid colours of the African garden; the butterflies and birds.

Behind the house was a round thatched hut where Nyagga and Mohammed lived; the kitchen was separate from the house itself too. One of the consequences of an external kitchen was that meals arrived cold to table during the rainy season – for the rest of her life, one of Dorothy's criteria for a meal was that it should be hot. Cooking was by a wood or charcoal stove and without refrigerators there was a limit to how much perishable food could be kept; some food came in tins – butter, vegetables – while sugar, flour and rice arrived in sacks. Soap came in long bars, over a foot long, and was of two kinds; made either from yellow coconut oil or soda. The latter was blue mottled and extremely hard, only used for washing really dirty clothes. Day to day goods were bought at the local Indian run *duka* – tea, matches – and some of these sold colourful bolts of fabric. The traditional skills of jam making and fruit bottling were essential. Dorothy was a practical woman well equipped to take over the household, and her years of Girl Guide camps were of course a good training for the basics of life in Africa.

But she had some skills to learn. While in England a generation earlier middle-class families would have had servants, by the early 1930s this would not necessarily have been the case, and Dorothy had to learn how to manage servants. Advice came in the form of *The Kenya Settlers' Cookery Book and Household Guide*, published by the Church of Scotland Woman's Guild, which included recipes and household management tips. If you did not like the soap from the *duka* there was a recipe for making your own. Mealie meal (corn flour) soap involved mixing 1lb caustic soda with 1.5 quarts of water, leaving it overnight and then mixing in 5lbs of mealie meal and 7lbs of melted pork fat. The mixture, when stirred would thicken like honey, and it would be poured into a shallow box lined with a damp cloth and cut into bars when dry. Advice covered destruction of insect pests (cockroaches, flies and ants); keeping chickens ('keep fowls if you want but keep good ones'); remedies and how to deal with poison (alcohol – administer any quantity of black coffee). There was a dictionary of Swahili words in common use and a number of orders to servants. These were strictly practical – 'see that the soup boils gently' or 'two guests are coming today'. Happily earlier instructions on how or when to beat your servant had disappeared by the early 1930s.

Immediate neighbours in the *boma* were the other government officers, a DO, medical officer and a teacher. The DO was now Lt Col. Anderson – there were many with a military background serving in the administration. There was a golf course nearby, as was typical, and later a small Anglican church – although it seems that there was not one in Kakamega in the mid-1930s. The present building of the Anglican Church in Kakamega,

Christchurch, was built in the late 1990s, on the site of an earlier church built in the 1950s, with a foundation stone laid by (then) Bishop Beecher. How long the Golden Hope served as church we do not know.

Social life for government servants revolved round the club with its tennis courts and billiard table. Owen immediately became involved in the church and church music; in most of his postings his musical life was centred around the church and its organ – he rapidly took over the role of organist and choirmaster – first in small communities, but eventually in the cathedrals of Mombasa and Nairobi. Owen's piano, the wedding present, was set up in the club house for musical entertaining. Also, most people, Owen included, owned wind up gramophones, and shared their record collections for musical evenings and dances at each other's houses. I still have some of Owen's records, 14 beautifully shiny shellac 78rpms stowed in a battered carrying box, and protected for 80 years. Mainly classical – Cesar Franck's Sonata for Violin and Piano, Bach cello sonatas, and extracts from opera (*Rigoletto* and *Lucia de Lammermoor*) but there are also waltzes and dance music. Besides music and sport – tennis, golf – bridge passed many hours. There were outings too – to the protected Kakamega forest for example, where there were cabins for a weekend retreat, to enjoy birdwatching and exotic orchids and butterflies, and the screeching and chattering of the monkeys. The life style – while sociable and enjoyable – was very far from the Happy Valley set. Sundowners yes, but they tended to stop at 7.30 in the evening. Dorothy made friends easily, and this was a period when they began to build up friendships lasting for the rest of their lives.

Dorothy also entered fully into the life of the *boma* with ease. Although in some ways a reserved woman, not given to showing strong emotions, she chatted and made friends easily and the community activities of the *boma* were not so different from those around the Methodist church of Beckenham; the League of Mercy, the East African Women's League (EAWL); organisation of fetes and dances, children's tea parties – exotic and ordinary at the same time, in the scents and sounds of Africa.

The records I have of this period of my parents' life are vague and anecdotal – only a few letters, as by the early 1930s the Wye alumni had more or less stopped writing to their alma mater, and there are only fragments of written testimony. There is a picture of Dorothy in an ox-drawn cart, clutching Boris, their dog, to her side. There are pictures of the garden – flowery and lovely, as was the English way – 'the English always built gardens'; Boris in the garden ineffectually chasing a cat up a tree; and pictures of the car being packed up for safari.

And their Visitors' Book. Life was mutually very supportive among the white community, and because distances were great and communications poor, people were expected to provide accommodation to colleagues as they travelled around from station to station. This is why the Visitors' Book provides such a good insight into their life. In the first year they were in Kakamega, they had three visits from Bill and Hilda Lyne-Watt, two from the Watsons and two from John and Stella Booth. John Booth, another Old Wye Man (OWM), had joined the Agricultural Department in 1931, and from 1932 to 1938 he was based in Kavirondo (Central and North Nyanza) where he was responsible for the Bukura Agricultural Training College and its experimental farms, which provided training for African extension workers. Tom and Margaret Watson also became long-term friends; also an OWM, he too was an agricultural officer in North Kavirondo (later known as North Nyanza). In fact in their first three-year tour, Owen and Dorothy had about 20 different sets of people staying with them, of whom at least half became lifelong friends – names I remember from childhood. Roger Lambert, an OWM who had switched from agriculture to the administration, was now based in Kakamega, 'down among the gold mines' as he wrote in a letter to his college.

Bill Lyne-Watt, the former BEADOC farmer, was by now senior agricultural officer responsible for Nyanza Province and based in Kisumu, the provincial headquarters. He was a man of fiery temper and a passionate golf player despite his war wound – he was one of a group called 'the Roaring Forties, who were well known on the Kisumu Golf Course for their highly coloured shirts – if not for their golf'. Bill was also a talented teacher and besides strengthening African subsistence agriculture, he instigated the cultivation of both cotton and rice on the shores of Lake Victoria. He had a very Scottish sense of humour and used this in his teaching – Bill and Owen worked closely together throughout the 1930s and were friends (and golfing partners) until Owen left Kenya in 1949.

He was also adept at drawing good people – not necessarily from Wye – into the service. One individual of lasting consequence to the service was Tom Hughes Rice. Tom, a lean wiry Welshman with an airman's moustache and a humourist's eye, had come to Kenya in 1928, at the age of 21, as a pupil farmer where he learned and managed a number of different types of farm – sugar estates, coffee and livestock. While Lyne-Watt was district agricultural officer of Nyanza Province he recruited Tom into the service from a commercial farming background. He would have seen in Tom something of his own background: a training in wider agricultural practices in Kenya as a grounding for constructive work in the African areas.

Outside of Western Kenya as well, Wye College continued to exert its influence. By the mid-1930s Stuart Gillett had become something of a coffee expert, with numerous specialist papers published. He worked first with settler farmers, but was later effective in developing African coffee growing in Meru and Kisii and was based at the Scott Laboratories where his job description defined him as an 'experimentalist'. Ted Gaddum was by now a cotton man, based in southern Nyanza and was later posted to Kilifi on the coast where, with his wife Dorothy and daughter, he worked on cotton and ground nuts, setting up an experimental farm with a training school for Africans. Windley, who joined the service at the same time as Owen, became a specialist in wattle research with several publications to his name.

For those based mainly on station, life followed a straightforward pattern. According to one officer, the daily rhythm was thus:

> rising at 6.30 after a cup of chai brought by the boy, breakfast at 7.15 on the verandah and get to the office at 8;

> knocking off 12.30 for lunch at home, back to the office by 2, then finish at about 4 tho' usually exceed these hours. After that, a quick cup of tea and golf or tennis if fine and squash if wet. Dinner would be any time between 7 and 11 – depending on other social events – mainly bridge. After dark the place is a mass of glow worms and fire flies.[15]

While Owen was in the office, Dorothy attended to household duties in the way of generations of women: overseeing the laundry, doing some cooking, sewing and making essentials. The sewing machine and patterns she had brought were much used. Sundays followed a routine of church, then drinks before lunch, then tennis or golf in the afternoon.

Extension work was at the heart of the agricultural officer's work, and officers were expected to spend at least one third of their time out and about on safari to the various communities in the district, or working on the experimental farms. The agricultural officers were in some ways the most visible face of the administration – known as the *bwana al shamba*. The image of the colonial administration in those days would be of the agricultural officer on safari, on foot with porters, by dhow, on donkey back, motorbike, horseback or car – whatever worked. In the earliest days, safari was on foot with porters – older hands talked nostalgically of how this kept officers in closer touch with the local people – a contact which was diluted as the car took over and roads improved.

One of my abiding memories of my parents' talk was their reminiscences of being on safari. Dorothy and Owen clearly loved safari – the packing up of the car, Boris the dog sitting in the car waiting to set off, the cooking pots and camp beds, the shower made from a canvas bucket. On safari they travelled with a cook and possibly a houseboy. In some places accommodation was a rest house, in others, tents and camps. *The Kenya Settlers' Cookery Book* had advice on what to take on safari too.

> The housewife who finds herself responsible for getting a safari together will find it best to prepare a careful list of

> everything necessary beforehand, and to check off the list as packing proceeds. A boy who has been on safari is a great help as he generally knows what is required, but even the best boy should be supervised. Careful well thought out packing makes a difference to the number of boxes required, and to the speed with which camp can be struck.... The cook can usually be trusted to look after his kitchen necessities... he will require clean petrol cans for heating water, and if there is no safari oven another tin for use as such; a bucket, a frying pan, a set of five or six *sufurias* fitting into one another, lids for these or sufficient tin plates to serve the purpose, bread or cake tins, washing up bowl, a mixing bowl, a strainer, kitchen knife, fork, spoons.... For provisions, much depends on the district but milk, eggs and fowls can usually be obtained while in some places small game and birds ensure a reasonable supply of meat... if vegetables are not obtainable they should be included in the stock of provisions.... fruit must not be forgotten, lemons and tinned peaches... a couple of *charpals* slung outside the car and kept filled with water will give a constant supply of cold water... and so on and so on.[16]

Details are minute – on the types of cheese, of biscuits (a tin of plain biscuits), and recipes for bread and bread substitutes. Then there is a boiled ham, baking powder, beef extract, camp bath, curry powder, a methylated spirit-stove, and a camp range. The length of a safari and the number of villages to visit varied, but often the officers would be out on the road for several weeks at a stretch. Good packing and planning were obviously very important.

Often they stayed in a local bungalow, or a chief's *banda* in the village. One photo shows Dorothy standing on the threshold of a typical camp *banda*, a long thatched building with very low eaves; she is calling Boris who is obediently, but with apparently studied slowness, strolling towards her. Behind the *banda* are some round thatched native huts, and standing alongside Dorothy is the local agricultural instructor and the village chief, smart in a jacket and jodhpurs. Nearby, looking on are two men and a young woman. Sometimes they would stay at a local mission station, and missionaries were regarded as their own good friends by agricultural officers, as they knew the local populations well, and were at the heart of African schooling and education. (There were sometimes mixed motives for mission hospitality; while offering shelter to passers-by is part of an ancient credo, having guests to stay earned the missions a special dispensation to serve alcohol in an otherwise alcohol-free regime.)[17]

But more often they were under canvas; if it was very hot, they might sleep under the stars, a mosquito net hung from a tree. Night noises included the screech of the hyrax, the shuffling of the buffalo; sometimes lion or elephant would be nearby. There was the clattering of birds before night fall, and again on waking in the morning. Chattering monkeys leapt from branch to branch. The night was bright with fire flies. Boris always went with them. The cook would look after the setting up of the camp and the preparation of food.

Travel was still far from easy by the early 1930s. Roads were often two tracks with a grass strip running down the centre; stretches were deeply rutted and pot holed. The journey between Kakamega and Kisumu – a mere 30 miles – might take an hour in the dry season, but a day in the rains. Any traveller took with him heavy car chains, rifles for keeping animals at bay and for hunting (for guinea fowl for the cooking pot), camp beds, tent, insect and water proof food boxes, water supplies, spare petrol and spare parts for the vehicles, folding chairs and tables, ground sheets and blankets. Chains were essential in the rains when earth roads could turn to quagmire and where the black cotton soils prevailed, all around the lower lands of Western Kenya, vehicles could find themselves buried to their wheel hubs. Owen described how once they had to traverse a river in high flow backwards, since the water backed up against the bonnet of the box body, and they needed to make sure the carburettor was downstream of the water flow.

If the journeys were challenging, so were the cars uncomfortable. Wooden tool boxes provided seats; but as Mary Hamilton[18] noted, 'it was no worse than a wagon and was at least quicker'. Wagons were still in use in the early 1930s for both transport and farming; one of the reasons that the South African settlers were welcomed is that they had a tradition of driving ox carts, trained both for transport and work. The oxen, harnessed for transport or ploughing, had an evocative charm and many were sad to see them pass. Michael Blundell, later minister for agriculture and negotiator for a multiracial Kenya in the 1950s and the run up to Independence, wrote lyrically of the delights of the ox span. In 1925, on his first morning as a young helper on a farm near Eldoret he wakes and remembers:

> The low toned bells of the Masai cattle were heard far across the valley, with the tinkling bells of the sheep and goats sounding clearly in the thin upland air. Smoke curled lazily from the beehive huts of the farm workers while below I could hear the sounds of the ox chains, the shouts of drivers and the occasional crack of the whip as the teams were being inspanned... a team of sixteen oxen, eight pairs one behind the other and linked by chains attached to the yokes, were often matched in colour: blue, roan, brindle, black, red, white or cream, and were an attractive sight... in front was the *mshika kamba* – a young boy who guided the leaders with reins tied across the horns. The driver, a king in his own right with a twelve-foot-long whip, the best of which were made of giraffe skin, would shout the names of the oxen... Sportsman, Michael, Nguvu, and crack the whip like a pistol shot. The leaders would straighten up and bend to the strain, the wheelers at the back would give a terrific heave and the eighteen foot wagon with its four inch gleaming iron banded wheels would slowly creak forward with four tons of maize or wheat.[19]

By 1930 cars and tractors had arrived, but oxen were used for ploughing in the Reserves well into the 1930s.

Officers met local chiefs and the local agricultural instructors to discuss programmes and developments with the LNCs. Sometimes the agricultural officers travelled with the local district administrators who made detailed reports on the state of the local markets, roads, health, economic activities and matters of law. There was a strong synergy and mutual respect between the administrators and the agriculturalists – all working together for the development of the area, of the African.

Once they had arrived at the *banda*, a *baraza* would be held where views were aired, complaints made and instructions given. Despite underlying tension over land in Kenya, at a local level the officers felt that there was a growing level of trust between the LNCs and the agriculturalists during the 1930s, as 'the Councils began taking a lively interest in agricultural services'.[20] The works in Nyanza focused on the opening of new cotton areas, some swamp reclamation, commencement of coffee growing and tobacco growing, extension of rice growing (around Kisumu) and – crucially – prevention of soil erosion.[21] African agricultural inspectors were the front line in training and discussing with local farmers, distribution of improved seeds, and advice on marketing and quality control, and without them the 'thin white line' of agricultural officers would have had much less impact. So while there was only one agricultural officer per district, about a hundred native agricultural inspectors were working in Nyanza Province in 1935, over half of whom were employed directly by the LNCs.

While meetings with chiefs and other dignitaries could be set up in advance through the network of agricultural instructors, the group discussion, the *baraza* was very important, and it was essential that the officers could lead a meeting effectively, with respect and intelligence. Speaking the vernacular helped, and skill in oratory was useful to transmit ideas, and get their perceptions across, and Owen's thespian and musical career in Nairobi was a fine background. Oratory in the *barazas* came easily to him – it was a lasting skill since when, in later years, he retired from Kenya and from farming and became an Anglican priest, his oratory showed from the pulpit. In largely illiterate societies this need for communication is self-evident although it is not clear whether it was among the skills taught at Wye.

Besides oratory and exhortation there was the practical matter of the distribution of improved seeds and demonstration farms, where the department was teaching soil conservation and the use of manure. There was also a demonstration smallholding (as distinct from experimental plots for different crops) at Bukura, the beginning of recognition that agricultural development in the native areas needed to integrate cash crops with food and traditional cropping – in Nyanza the cash crops were cotton and maize.

The third strand of work in the Agricultural Department was in marketing and crop quality improvement. Cotton and tobacco were grown for export rather than subsistence and the quality had to be good and consistent – and the same applied to maize when there was surplus above and beyond household needs. Each district had its own priorities, needs and climate. In Kakamega, for example, there was a strong demand for timber from the mining industry. Indian traders played an important role in marketing crops, and in central Kavirondo there was a voluntary system of central inspection, actually paid for by the Indian traders. By 1935 there were about five ginneries, probably owned by the Lakha dynasty, an influential Ismaili family, producing 12,000 bales of cotton. The ginneries were also used for oil production and milling *posho*. In fact this latter innovation had a profound impact on women's lives. *Posho* is maize flour, traditionally ground fresh for each meal by womenfolk in the individual household. But with the advent of the *posho* mills women could carry a headload of maize to the nearest mill or ginnery and have it ground there, a considerable saving of labour.[22]

Around this time Owen renewed his schoolboy friendship with Leonard Beecher, who was now a missionary and priest. Beecher had strongly opposed the early draftings of the Native Land Trust Bill, and became a prominent adviser on Kikuyu politics. Leonard and Gladys and their children lived very simply at a CMS mission station at Kahuhia, at the foot of the Aberdare Mountains. During this time Leonard was working with the local population and encouraging new economic activity, including the establishment of farms; he set up a Cooperative store, and demonstrated mixed farming through his own smallholding. He was fluent in Kikuyu; he had a special palate made for his mouth so he could speak Kikuyu like a Kikuyu. Gladys too spoke fluently – but in a rather formal style. At their golden wedding celebration in Kabete church a Kikuyu lady was heard to say of Gladys 'that European lady speaks wonderful Kikuyu. But like my grandmother'. Leonard also established a school for teaching English, was a scholar in both Swahili and Kikuyu, and wrote, with Gladys, a Kikuyu-English dictionary and produced a Swahili bible.

He became actively involved in the coffee disputes, supporting a Kikuyu teacher named Parmenas who had challenged the Native Grown Coffee Rule which had forbidden coffee planting in the Kikuyu districts, despite the strong wishes of the local African farmers. These demands had been fiercely resisted by settlers, partly because they wished to protect their market domination, but also to protect their high standards of husbandry and disease controls, driven by a not wholly unjustified wish to protect the reputation for quality which might otherwise be undermined by infection from poorly managed smallholder coffee plantings. In the event, the settler wishes prevailed, and indeed, when coffee was finally officially permitted, it was not in Kikuyu districts – it was in Western Kenya, far from the Kikuyu.

Beyond the narrower confines of agriculture, Beecher saw the importance of political representation for Africans, although, as he wrote in June 1935:

> as a missionary I try to keep politics as much in the background as possible. There are times however when it bursts upon one's daily work. Men like Parmenas should be permanent members of the Race Relations Committee, and the members of Legco representing Native Interests should have a permanent advisory council of representative Africans whose opinion they would take and follow in all matters representing African Interests. At the present moment, the people here and elsewhere have absolutely no confidence in these Members.[23]

Later Beecher wrote an important report on education for Africans[24] – he proposed universal primary education, a far sighted objective which was criticised both by some of the settler community who did not particularly wish to see an educated African population (and workforce) – and by the Kikuyu who argued that universal primary education was an inadequate aim, and that full secondary education was needed.

Close to Beecher, Owen became friendly with Canon Leakey. He was also a friend of Archdeacon Owen – a photograph shows them together outside the tiny church in Kisii a few years later. The activist role of the missionaries and the technical, practical, support and training among the agricultural officers made a potentially powerful tool for supporting African development, reflecting perhaps the robust Christianity Oscar Watkins described when he first arrived in Kenya.

After the departure of Grigg in 1932, the Colonial Office pressed for an immediate increase in African export production, and the new governor, Sir Joseph Byrne (1931–37) was less influenced by the settlers than his predecessors.[25] Throughout the 1930s, there was a consistent if modest sum set aside specifically for native agriculture amounting to about 20–24 per cent of the total agricultural budget and, in addition to this, the resources for the Scott Laboratories and for inspections and grading would have directly benefited African farmers. It was under Byrne that important changes were introduced, notably in the growing of coffee. In 1934, after continuing pressure from many officials (the DC in Kisii, Clarence Buxton, had been very active), the Governor's Office finally authorised experimental coffee schemes in Kisii, Embu and Meru. Nurseries were established in each area, but 'an undertaking was still given to the settler Coffee Board that planting would be limited to a negligible acreage in districts remote from European plantations, and would only be increased after consultation with the Board'.[26] The coffee controversy exemplifies quite neatly the tensions between the coffee growing settlers and their cosy relationship with the governor, and the drive for local development felt by officials such as Buxton who were working in the Native Reserves and wanted to see livelihoods strengthened and the local economy modernised.

The settler pressure would have been hard to resist. In the 1930s, Kenya comprised a white colony of some two thousand rural landholdings, a European urban community making up, with the rural settlers, a white population of some 5,000 souls, an Asian group of some 20,000, and an indigenous African population of about 3.5 million. While the African population – and even the Indian – vastly exceeded the number of Europeans, effective power in the colony lay in Legco, a committee of mainly white men with one Indian and one Arab member plus a European to represent African rights, and a number of boards – such as the Coffee Board – and committees which could exercise a strong white control over policy. As Liversage, an economist appointed to the Agricultural Department in 1930 writes, 'the system formed an admirable vehicle for indoctrinating the government with the immigrant point of view'. Along with various boards and committees:

> the official members of the administration have the settler point of view presented to them at breakfast with the morning paper, deputations and private interviews during the day, at sundowner parties and in clubs in the evening and the process of conditioning is continued at committee meetings. One sided repetition whittles away doubts as water wears away a stone.[27]

and he argued that a consequence of this pressure and bias is that the settlers massively held back the development of African agriculture.

One curious anomaly of the department's activities was that, while it had established a number of training establishments for African agricultural instructors and extension workers by the mid-1930s, at Bukura and the Scott Laboratories it had not provided much in the way of training for the European farmers, even though the Director of Agriculture was concerned that:

> a considerable number of farmers possessed no previous experience when they settled in the colony, and it stands to the credit of many of them that they have learned the 'art of

> agriculture' and have achieved success, but there are many who apparently do not have a 'bent' for farming.[28]

It took the unexpected vision of Lord Maurice Egerton, rather than the department, to redress this imbalance. The youngest son of the Egerton scions of Tatton in Cheshire, Lord Egerton arrived in Africa in 1926 and acquired 26,000 acres around Njoro. He recognised the problems within European agriculture and, in 1939, donated 500 acres of his land for an experimental farm as the basis for a college specifically for white farmers (and their wives). John Booth, another close friend of Owen's (and OWM), who had spent several years as head of Bukura Training College and Experimental Farm, became the first principal of Egerton College.

Egerton College has now developed into Egerton University, covering a full syllabus but with a strong continuing focus on agriculture, and Egerton himself is seen as a visionary by the professors at the university; he was certainly eccentric. On his great estate, he built three houses, and the final one is a small castle dating from 1938. The castle was designed with effect in mind. Set in landscaped gardens on the brow of a hill with views all around across the Rift Valley to the Mau, the castle was a mini replica of the ancestral pile at Tatton, even to the wide, long avenue of trees through which the castle is approached. The roof tiles were of stone and so heavy that the roof trusses were built of oak imported from the UK as Egerton believed African woods would not be strong enough. The panelling in the great hall was made from timber imported from Uganda, sent to the UK to be fashioned and re-exported in kit form. He installed a huge Aeolian organ, also built in the UK; he could not, however, play it – it had paper rolls for music. It could be heard from four miles away.

In 1935, after a year in Kakamega, Owen was transferred to Kisumu where he took over from Bill Lyne-Watt as Senior Agricultural Officer for Nyanza Province, on a temporary promotion while Lyne-Watt went on leave. He and Dorothy loved Kisumu; he wrote to contemporaries from Wye:

> The most enjoyable part of our tour we spent in Kisumu in 1936. It's an excellent town with a rapidly growing volume of business, mainly owing to the increased prosperity of the large Native Reserves of the Nyanza Province of which Kisumu is the natural commercial centre, and to the mining activities thereabouts. It is a pleasantly warm healthy town (more or less healthy that is to say) with a most excellent crowd of people who go out of their way to make the newcomer welcome.[29]

It seemed to be an exceptionally hospitable place, with a strong camaraderie among the women; some of the women's groups such as the EAWL were particularly active in Kisumu. Bill's wife Hilda wrote that 'The spirit of hospitality extended to all and sundry by the womenfolk of Kisumu has warmed the heart of many a stranger passing through'. In a local newsletter she writes of the warm characters in Kisumu, where the women's influence has made itself felt – from the earliest days of settlement. 'No appeal to the women for help has ever been made in vain.' There was clearly a special atmosphere in the place which Owen and Dorothy felt.

It is still a pleasant town: streets lined with flame trees and many buildings still standing that Owen and Dorothy would recognise – the Nyanza Picture House, now a Redeemed Gospel Church, the Mistrinarshilaxman Bros emporium; and the clock tower which had been erected in memory of Kasim Lakha, an early Ismaili immigrant to East Africa.

Kasim had migrated towards East Africa with a few Ismailis following a dreadful drought in the Kutch in India. Dressed in rags, he landed in Zanzibar in 1871 and then moved, with his growing family, to Lamu in 1877 and then on to Mombasa in 1903 once the Uganda railway was open. He was a successful businessman and seems to have invested in both Kenya and Uganda for in the end he died in Kampala in 1926. He left behind four sons, who together built extensive business and industrial interests in projects such as a hotel chain in Uganda, coffee curing works, coffee plantations, and traded in hides and skins and sisal. In Kenya, they were among the early pioneers in cotton ginning and the oldest was the founding chairman of the Oceanic Hotel Group of Companies in Mombasa, and it is he who is held as a 'Hero of the Ismailis'.[30] The memorial stone in the clock tower was laid by the then governor, Sir Robert Brooke-Popham, showing recognition for the Ismaili community and their business contribution to East Africa.

In the mid-1930s Kisumu was a pleasing town, with industry, hotels, fishing – a rich life. In 1938 there was still no networked electricity, but the East Africa Industries Company was said to be thinking about it. There were two decent hotels – the Kisumu Hotel run by J. R. Maxwell, and the Marine, run by the McClures. There were two companies offering car hire and taxi services 'operating over any distance', and an excellent road to Kakamega. There was a daily mail service to Kisii and beyond, carrying African passengers, and a regular bus service carrying Europeans between Kitale and Kisumu via Eldoret and Kakamega on Tuesdays and Fridays, operated by Roadways Ltd. Kisumu was something of a transport hub, being on the lake with direct communications with Uganda and the railway through to Mombasa. As early as 1926 there was a scheduled sea plane service between Khartoum and Kisumu. Flights were offered about four times monthly. The journey took two days, and the plane stopped at Jinja and Butiabe (for lunch), then at Mogalia and overnight at Shambe. On the second day, the first stop was at Malakal, then Kosti (for lunch) arriving at Khartoum at 2.30pm. The fare was around £65 each way.[31] Within the year the service was extended to Cairo. There were Empire and Sunderland flying boats alighting and taking off in the waters of the gulf, just below Kisumu, as well as a service for small ports of the gulf, a motor boat called the Kazi Moto which pulled a string of lighters full of cargo and/or people.

Shaped by the lake and Homa Bay the form of Kisumu has not changed since the 1930s. The Golf Club lies across the bay from the town, through the heart of which runs the great railway. Today it is operating again for passengers, thrice weekly overnight connections to Nairobi ('The Florence Express'), a memorial to Florence Preston, after whom Kisumu was first named (Port Florence). She must have been tough; she was with her husband through the whole of the construction saga and stood by his side, driving in the last key of the railway, when it was finally opened in 1903. Another service runs four times a week to Butere, and this service is known as the Butere Dove. Links between Uganda and Kenya across the lake were important in the 1930s, and much stronger than today, when much of the lake services have been replaced by air. But one of the curious legacies of the British Government was to give to Kisumu, as a leaving present at Independence, a roll on–roll off four abreast rail ferry. But as nobody ever matched the investment at the other end, the Uganda side, it only operated for a few years. And by the time of Independence the technology was already outmoded. It sits in the port area, under the careful eye of the Port Authority and is an attraction for school parties. What, one wonders, are they taught?

In the *boma*, several government buildings and many of the government houses still stand. Even the house Owen and Dorothy lived in is still standing and occupied, one of three sitting beside what is now Jomo Kenyatta Highway, completely recognisable from a photograph showing Dorothy outside the entrance and covered verandah. The garden they had inherited from the Lyne-Watts for this temporary promotion was well planted – Hilda herself was a fine gardener, and her vegetables won prizes in the annual Nairobi Horticultural Society Show.

The highway leads out to the Nyanza Club, which was very active when they were there. In the reception area of the club there still hang lists of all the presidents, dating back to 1915, when the first president was Hastings H. Horne, the brother of Edward Horne, the first DC of Meru (Edward because of his diminutive height was known as Short Horne, his brother as Long Horne). When Owen and Dorothy lived there the president of the club was Sydney Fazan, the local provincial commissioner and founder of the Kenya Kangoni Cricket Club. The furniture in the Nyanza Club today looks as if it too dates from the 1930s, and the dining room is dark, as the interiors of all the clubs seem to be. Is this a legacy of the early white anxiety about actinic rays? The club was of course built on a prime site, overlooking the bay; it is now dwarfed by large mansions which belong to successful politicians – it is still the prime site where those with money and influence choose to build.

As ever, white social life revolved around the club with a swimming pool in front of it and a library and occasional dances. There was a Yacht Club and a parade ground in town which could be used for hockey or cricket, the latter using a matting wicket. In the evening people could drive or stroll along the lakeside road, with a good chance of seeing hippos coming on shore to browse on the grass. The bar at the Kisumu Hotel was also a favourite resort for younger people, the scent of frangipani mixed with the smell of bat dung – the corrugated iron roofs were infested with bats.[32]

The other centre of Owen's life was of course the church. The first church built in Kisumu was on the site of what it now St Stephens which has a reputation as having the best choir in Kenya after All Saints Cathedral in Nairobi. Indeed the church looks well cared for with the electric organ with all sorts of stops in such good condition that it is clearly well used. There is also a harmonium which dates from the 1930s. But the main Anglican church (the *wazungo* church) in the 1930s was St Peter's, an imposing stone building, lofty with fine timber trusses and corrugated iron roof, and a pipe organ. It is now a centre for aid projects – when I visited it in 2010, the nave was full of cardboard boxes piled high containing training programmes and papers; and the chancel was jammed with brand new bicycles, still in their dusty plastic packaging, labelled with their destinations but not as yet delivered. The old pipe organ is broken and unused now, but in Owen's day it was a fine instrument, which he played when he took over as choir master and organist.

On the west wall can be seen a history of early settlers, of their death and struggles: engineers, a builder of churches, a pioneer padre. One plaque reads: 'To the Glory of and in memory of the Officers and Engineers of the Uganda Railway Marine, Erected by Brother Officers and Engineers'; now all but illegible, the tarnished brass plaques list 20 or 30 names: 'Name, Rank, Date of Joining, Date of Death, Place of Death, Age'. Another reads: 'To the glory of God, in fond and loving memory of E. M. Boughton Knight of Downton Castle Ludlow, England, aged 29. He died of fever at Karunga and was buried in the white man's graveyard at Kisumu, February 1st, 1904; Fervent prayer availeth much' (but did it?). And yet another; 'To the glory of God and to the memory of Rev F. H. Wright, CMS 1895–1925, Chaplain Kisumu 1913–1925 – A Pioneer Padre'.

At the end of 1936 Bill Lyne-Watt returned from leave and Owen himself returned to Kakamega for the remainder of this tour. He was commended for his work at this time by Provincial Commissioner Fazan, who wrote that he:

> tackled the job of Snr AO during Lyne-Watt's absence very thoroughly and... achieved valid results. I think... policy and programme have grown a good deal clearer as a result of our various discussions together and recommendations to headquarters, and they are being translated satisfactorily into action.[33]

Fazan was an important thinker about agricultural policy in Kenya, and he and Owen evidently had a strong working relationship. When Owen departed on leave he wrote:

> Before you go on leave I take this opportunity of thanking you most sincerely for all your good and zealous work on behalf of agriculture and the advancement of natives in this province. You have a lot to look back on with satisfaction in the way of results achieved in this tour.[34]

New crops, new markets; improved husbandry and a policy and programmes aimed to improve the life and welfare of the African. The discussions were the fruit of safaris and *barazas*, in the communities and villages, discussions with LNCs, and local leaders; working closely with the district officers and often going on safari with them: the notion of 'trusteeship' in action. But the early 1930s were a frustrating period for the agriculturalists; there was a strong will to improve African agriculture and this is seen in many of the Agricultural Department reports and writings by the officers themselves. But they were severely hampered by lack of resources; even by the end of the 1930s, there was only one agricultural officer and perhaps six or eight trained African instructors per district. No wonder they found it difficult to achieve a lasting impact despite the willingness of farmers to try new methods. At the same time, the department was putting £700,000 into research into pyrethrum, coffee, wheat and cattle. These researches came to benefit Kenyan agriculture in the future, but at this time the benefits fell to the European farmers.

After their return to Kakamega, Dorothy found she was pregnant to their great delight and they planned their home leave so as to return to the UK for the birth. There were always health hazards in Kenya – malaria was (and still is) widely endemic and most officers, Owen included, succumbed several times. Communications were slow, and the nearest decent hospital was in Nairobi. It was therefore normal practice for officials' wives to return to the UK to have their babies, and this is what Dorothy did. She and Owen travelled by boat from Mombasa, timing their home leave for the confinement and recovery. But when she arrived in the UK she was suffering from high blood pressure and the baby was still-born at 20 weeks. It must have been a terribly sad and distressing time for them and for their families at home who would have been thrilled at the news of their first grandchild. Owen's career was going strongly, but over the next ten years Dorothy lost three more babies, each at about five months. Not only was it a very sad home leave, but – and this they would not have known of course – it was to be their last for many years.

In their second tour in Nyanza Province, when they returned from home leave, they were posted for a few months to Kitale, near Mount Elgon and the Uganda border. It was the first time that Owen had been posted to a district that mixed African agriculture with European farmers. The setting was – and still is – very beautiful. From the verandah of the club (with a splendid rolling 18 hole golf course) lay open country to Mount Elgon. Their house was again a classic government bungalow, with two bedrooms and a sitting room opening off a deep brick pillared verandah; and corrugated iron roof. The kitchen was integral to the house, and the water supply came in large jerry cans drawn on donkey carts; the water was heated by a wood fire placed beneath a water drum in a brick built fireplace. (This method for heating water is still used – at the Eldoret Club for example, there is a fine structure of large cans over an open fire.) There was music in Kitale – the pipe organ in the Anglican church, St Luke's, had a very sweet and true tone, and again Owen took over as choir master and organist. This organ is still in good condition, the pipes carefully painted in yellow and blue. The church, now promoted to a cathedral, stands in gardens with monkeys playing in the jacaranda trees.

Dorothy was pregnant again, and they agreed not to try to make for England but to stay safely in Kenya. Kitale had a nursing home, a simple affair with four bedrooms although no water or electricity. But once again she lost the baby – a boy – at about five months. Part of her Methodist upbringing was a stoicism which must have been a support; that and faith. There could have been no reason to expect, when she left London for Africa, that she and Owen would not be able to start a family as soon as they were ready. Friends around them were of an age: Michael and Dilys Grieve, Tom Watson and Kay, John and Stella Booth, all their close friends in Kenya were starting families; it was painful for Dorothy.

For the rest of her life she blamed herself for leaving for England in the first pregnancy, believing that the strain of the journey had caused a weakness in her which would last for many years. But looking at her records and comments from a doctor later, after yet another failed pregnancy, it seems that her condition may have been a combination of factors, none attributable to the voyage, but more the result of lack of understanding of certain medical conditions – understanding of which has much developed in the past 50 years – perhaps pre-eclampsia, perhaps an autoimmune condition or perhaps an x-linked recessive genetic condition. Her lost foetuses were all males and her only successful birth was – clearly – female. There is really no way of determining the causes now, but her regrets and eternal feelings of guilt at the voyage home the first time were almost certainly misplaced.

The next few years would see many changes as the war took over and Kenya's mission was now to feed the armies of the front lines.

5

THE WAR YEARS

> 'The defence of Kenya must rank after the defence of the Soudan. There should be time after the crisis in Egypt and in the Soudan to reinforce Kenya by sea and rail before any large Italian expedition can reach to Tana River. We can always reinforce Kenya faster than Italy can pass troops thither from Abyssinia or Italian Somaliland.'
>
> Churchill, 1940

From Kitale, Dorothy and Owen moved on to Kisii, towards the end of 1938. Kisii was a very attractive posting, a small green *boma*, 5,000 feet up in the hills. The main government offices were huts with thatched roofs, although the Department of Agriculture was in a little brick built single storied building with corrugated iron roof. Their house was a standard bungalow but surrounded by a pretty verandah and set in a large garden with beautiful mature trees – avocados, mangos, pawpaw and eucalypts – all round. They kept chickens, had a cat and kittens, and the faithful Boris still held the fort. I still have the beaded Gusii wooden stool that Dorothy bought there, and there is a photograph of her proudly displaying a hardwood coffee table which has always been part of my life and sits now in my study laden with papers for this book. But not everyone had good accommodation – when Tom Hughes Rice arrived to take up his first post in Kisii, there was apparently no house for him, so he had to make do with 'a little office at the Seed Farm and a bay of the implements shed with papyrus matting – no light, no running water and no WC!' 'However', he added philosophically, 'I was mostly on safari so it was no great hardship'.

When they moved there, Donald Storrs-Fox was the DC. Donald had arrived in Kenya in 1920 as an assistant district commissioner, and had acquired a reputation as a formidable hunter of elephant, rhino and lion. While Owen evidently had a good relationship with him, Donald, a man of strong convictions, was at perpetual 'war with "Sydney Herbert" Fazan', according to letters from his wife Betty. Donald had first arrived in 1920 and he was posted to Kakamega when it became the new provincial centre; his views are from the first uncompromising. He writes in an early letter home:[1] 'The European settlers as a class are out to exploit the nigger [sic], settlers pay 5–7 rupees a month and do not look after housing etc.' He goes on, 'There are three groups, settlers, missionaries, and officials. The officials are pro-white or pro-native'. Most that he has met are pro-native as they want to see fair play. 'In the reserves', he goes on, 'this is not hard but in the settled areas they are less pro-native as it is made so jolly unpleasant by the settlers'. Illuminatingly he then writes in another letter: 'We've not met many missionaries, but they are disliked and distrusted by both the other parties – being suspected of extreme pro-nativism tending towards sedition' (but he adds 'keep this in the family as it may be deemed too disloyal').

Donald's next letter home provides another fascinating insight into perceptions of the time:

> I'm not sure I didn't give you the impression in my last letter that the nigger was very downtrodden in this country. Well, his condition isn't so very bad; there is a lot to be thankful for, and *he'd be much worse off under any race other than the British*. [My italics][2]

He went on to become increasingly articulate about various issues of native rights, opposing forced labour and requisition of foods during the war, as we shall see later.

It is undoubtedly the case that the settlers were suspicious of the missionary efforts at education for the Africans – seeing, rightly, that this would be the path towards development and economic freedom. However, it is not true that settlers never looked after the interests of the African. Many farming wives provided some education and usually a clinic and basic medical services (one wife writes how 'I shall never the forget the faces of some friends who arrived for a drink on Sunday night... to find me sitting in an engrossed fashion stitching up the tractor driver's leg with black thread and a darning needle').

Other officials in Kisii included Roger Lambert, an old friend from Wye, as assistant district commissioner; an MO (Bloomfield); revenue officers and police; and Tom Hughes Rice. Dorothy and Owen became particularly friendly with the Storrs-Foxes and with the eloquent Tom Hughes Rice and his wife Margaret; in his manuscript memoirs Tom recalled meeting Owen 'who I liked and admired and who with his wife Dorothy remained dear friends' there. (Writing as he was in 1977 he added that he still kept in touch with Dorothy – Owen had died in 1972.)

Lambert welcomed the arrival of a Wye man as agricultural officer. He wrote to Wye College with pride and colour of the achievements of the administration in Kisii in the mid-1930s:

> Our 'flock' consists of some 350,000 Kisii, Jaluo, Watende and Abbasuba. The last named are a handful of people who live on islands and along the lake shore, but it is good fun visiting them by canoe. The Kisii are Bantu, and their country is more likeable than they are themselves. They are lucky in having such fertile country where the rains seldom fail... they raise most of their taxes by working on the neighbouring tea plantations which are very flourishing concerns. The Luo are Nilotic people who live by the lake shore. They are cheerful fellows, pretty nit-witted, but very progressive these days. We are not much troubled by the politically-minded gentlemen, and one feels that the average native does appreciate that we are here to improve his lot. They were pretty hard hit by the Depression, but, thanks to our getting them to plant cotton, ground nuts and sim-sim, they are now comparatively prosperous. We boast an agricultural officer who, strange to say, is not a Wye man [that would be remedied when Owen moved for a spell to

> Kisii]... we export wood, quantities of hides and ghee, not to mention cattle on the hoof. I have quite a lot to do with it, mainly rubbing in the preaching of the experts, organizing marketing facilities and punishing the idlers and so on. Last year we organized a Provincial Agricultural Show on a big scale. It went off very well and HE, who came up for it, was delighted. What was intended to be a parade of winning fat-stock ended in a mounted charge past HE, most of the riders being bedecked with ostrich plume. It was much more impressive than what we had arranged![3]

Tom Hughes Rice was a colourful character; he had a Welsh gift of oratory as well as eloquence in local languages. On his first day of service, after being recruited to the service by Bill Lyne-Watt from the commercial farming sector, he attended a *baraza* with Lyne-Watt (who was known for his fiery temper) and another colleague. He wrote in his memoirs:

> I had reported to Bill Lyne-Watt in the morning and that day the PVO [provincial veterinary officer] Mr Emerson and he were to attend a *baraza* near Kibos to hear the protests of the Luo Cotton Growers' Association of the Kano plains, against the heavy handed methods of the assistant agricultural officer in charge of central Nyanza district. It was clear from the outset that we were in for a stormy session. An ex agricultural instructor, named John Paul – who was, I think, more than a little 'off his trolley' – had already worked the audience of about 150 people into a fine state of shouting. Lyne-Watt was not at all tactful and fell at once into a slanging match with John Paul. They stood furiously face to face and then Bill unwisely took hold of John Paul and shook him to and fro. This really set the party alight, and in one second the audience had surged round us and was starting to rough us up. Now comes my part in the event! I have a very strong carrying voice, and I opened up with great clarity in Dholuo [the Luo language] saying: '*Linguru jowa! Mos, mos! Utimuru nade joduong! Weuru koko! Udwage negowa! Beduru piny owinjore wach.*' Or 'Be quiet for dear sakes!

> Gently gently! What are you doing gentlemen? Stop this noise! Do you want to kill us?'[4]

This little speech in local dialect drew the attention of the crowd, and Tom was then able to talk quietly with the men and listen to them; he promised he would pass on their concerns, and they then discussed quietly for two or three hours. From a situation of incipient violence, Tom managed to transform the mood so that they ended up all shaking hands – and he learned the importance of listening, oratory, and linguistic skill. (As a side note, Tom had been working on various farms for some years and had learned a 'fluent but non-grammatical' – his words – version of both Swahili and Dholuo; when he came across Stuart Gillett who was advising on coffee growing, he began to think about joining the service not least because of the opportunity to learn local languages to a higher and more formal level – since settlers so often did not speak local languages well.)

Another time he was less successful: he did not speak Masai, and needed to work with an interpreter. On one occasion, he was addressing a *baraza* of Masai, and came to a tremendous peroration about the dangers of overstocking:

> Deserts are forming hour by hour, your springs are drying up, soon your cattle will waste away as they search desperately for drops of moisture among the rocks. Are future generations to say – look, that was the country of the Masai, the great warrior nation, the scourge of lesser breeds? Because they lacked the sense to preserve their own heritage they have dwindled to a few poor families who beg a meal from the Kikuyu and Luo, now the rich and strong. There is not a day to lose; warriors awake; delay no longer![5]

Pointing dramatically to his watch, tapping his left wrist for emphasis he cried with passion: 'I say to you, time is against you; the sands of time are running out!' Amid a rapt silence, the interpreter arose to render the great speech into Masai. The message was short and blunt: 'The *bwana* says his watch has stopped'.

Kisii district rose from the shore line of Lake Victoria to high country, which would later become rolling tea estates. The officers going on safari visited the islands on Lake Victoria, and Tom writes of his work with the islands of Businga and Mfangano, and lake locations at Gom, Kasiguna, Kaksingiriri and Gwasi. He writes:

> I used to set off in the LNC dhow and sail all along the coast, calling at settlements adjacent to the shore. I finished my safari of usually a fortnight at a time with a foot-safari to be met by the LNC lorry at the nearest road as there were no motorable [sic] roads through the lake shore locations then.[6]

On a later Rusinga safari, they were dismasted in a sudden squall, and were 36 hours on the lake before they could work their way ashore – after that the fisheries officer (then a Mr D'Eath) supplied vessels for the island safaris.

Owen and Dorothy undertook various island safaris – a blurry picture shows Dorothy with Owen, pipe in mouth, solar topee on his head, being rowed in a wooden dhow by an African in a tarbush, then being pulled ashore by an animated group of naked boys. Another shows Owen being rowed ashore at Kasigunga for a *baraza*, and one shows Dorothy sitting with the Commander on the stern of the fisheries launch 'somewhere on the Lake' – presumably sometime after Tom's mishap. There was a demonstration farm in Kisii, where among other things Owen was involved in trying to breed native cattle – the beautiful hump backed zebu – with imported strains to achieve higher yields; there is a series of photographs of Owen with three calves sired by a magnificent bull called Peter. This is particularly interesting since the work undertaken in this period to strengthen and cross breed local zebu cattle has been largely forgotten.[7] But the most worrying concerns were land erosion and flooding. Owen's photographs show a 'wash away' on the Kano plain and severe soil erosion in Wangaine and Suassi camp; by the early 1940s this had become the most important issue for African agriculture.

Island safaris, the successful breeding programme at the farm, animals and good friends, bridge and music. This was a happy place, far from the centre of events, but even here they were painfully aware that the clouds of war were gathering on the horizon. Communications were not bad for the period, but as war approached there would have been a grave feeling of distance from those in Europe – and as the war deepened, a strong sense of isolation and darkness.

Kenya was well away from the principal theatre of war in 1939, which was the battle for mainland Europe. But it was to play a significant role in the North African theatre the following year, and later was an important support to the Allies in the Pacific front.

As soon as war was declared in September 1939, censorship was imposed on all postal and telegraphic communication, petrol was rationed, maximum prices of food were set (and published) and Imperial Airways ceased carrying passengers. There was a sizeable population of Germans in Kenya, some 700, and they were rounded up and interned. Some managed to get away to Italian Somalia, or had managed to get on board the *Watussi* which sailed from Mombasa shortly before the war, although it did not benefit them much – the vessel was eventually scuttled by her crew just off Cape Point, when challenged by naval patrols.

A possibly apocryphal story is told about the agent in Mombasa for the German shipping line. He was a popular man with all the communities and kept a remarkable cellar. During the Munich crisis in September 1938 he felt that war was inevitable and considered it his duty, as a good German to prevent 'valuable munitions of war from falling into enemy hands'. The story goes[8] that all the Germans in Mombasa were directed to show their patriotism by disposing immediately of the contents of the agent's cellar before hostilities broke out. At 'great personal inconvenience' this race against the clock was won, but the sense of wellbeing brought about by a surfeit of Liebfraumilch, and the sense of a duty well carried out was shattered by the news that agreement had been reached and that war was 'off' for the time being.

All able-bodied British white men were called up to join the Kenya Regiment. The aim of the new regiment was to train up junior leaders to cope with the increase in numbers of African soldiers in the King's African Rifles. Of the 9,000 white Kenya males, nearly half served in the forces; there were 3,041 men in essential occupations – Owen was one of these – and the balance were either too old or in other ways unfit. Preparations for war were made. The workshops of the Railway in Nairobi which were the only large well-equipped mechanical workshops in East Africa, and normally used for maintenance of the rolling stock and harbour equipment, were now busy building armoured cars by cladding three-ton Ford lorries with mild steel plate; making mortars out of drain piping, and constructing mortar bomb casings; manufacturing anti-tank mines, gun lorries for the East African Light Battery and wireless vans. They straightened defective blades on propellers for the Royal Air Force and much more. In the last six months of 1940, they designed and built bodies for 250 ambulances, manufactured 72 three-inch mortars, 25,000 pickets for barbed wire, 600 four gallon water tanks, as well as anti-tank mines and hundreds of stretchers and other tools, and repaired scores of Bren guns.[9] In fact, after the fall of France, work for the East Africa Command had complete priority, and in the last six months of 1940 more than half the workshop capacity was devoted to equipping the forces.

Because of their importance in keeping the mechanics of the war going the railway buildings themselves were camouflaged, as Mary Hamilton recounts:

> I was supplied with eight Africans and an unlimited amount of paint for the interesting job of camouflaging the railway buildings in Nairobi. First I would look over each building, taking into consideration its surroundings and then draw up on paper a camouflage plan. A plan to hopefully blend them with their background countryside... While the Africans crawled about over the roof, carrying out my colour scheme, I stood on the ground shouting out directions.... After my eight Africans and I had finished painting the buildings in drab greys, fawns and greens I was

> taken up in a light aircraft to view my handiwork from above.[10]

The war really began to impact directly on Kenya when Italy joined Germany and declared war at midnight on 10 June 1940. At this point the entire Italian population was immediately interned – except for the apparently charming Furio who worked for Kitale Motors, and who had constructed the first and only air raid siren on the roof of the local hotel; he was also the only one who knew how to operate it, so he was allowed to stay free under a special guard.[11]

By this stage in the war – after nine months of active hostilities – it was clear that Britain was pretty much alone. France had fallen, the evacuation of Dunkirk had demoralised the British. The Americans were not yet engaged. Churchill believed that the Mediterranean was under threat and indeed that if the Italians launched a comprehensive offensive they could take the Red Sea coast and the Horn of Africa in a matter of weeks. Not only was the Mediterranean under threat but so were the transports and merchantmen in the Mediterranean because of the tremendous loss of escort vessels in the flight from Dunkirk. Britain needed to retain control of the Suez Canal and the Red Sea, so as to secure shipping routes to India. The strategic necessity therefore was to protect Egypt.

The balance of power in June 1940 was entirely in favour of Italian forces; the bulk of the British strength under Middle East Commander General Wavell was defending the Suez Canal with forces in Egypt. Elsewhere there were small British and French forces in the British Somaliland and French Somalia (Djibouti) but, since the French armistice with Germany, Churchill no longer trusted the French. When the Italians had entered the war, the RAF flew a couple of sorties over Abyssinia but without any force behind it – it was only posturing.

At the time, the Italians had some 300,000 troops in Abyssinia, Eritrea and Somaliland; well-equipped men with 400 guns and 200 aircraft, compared with Kenya's 8,500 men. But apparently the Italian forces were not the greatest fighting force; they were primarily native units, 'recruited on a tribal basis, not designed to fight conventional wars, lacked mobility, were terrified of artillery and were led by aloof Italians on horseback'.[12]

The Italians also had very poor intelligence.[13] They believed that Wavell had 100,000 men in Egypt whereas in reality he had 36,000 in Egypt and 27,500 in Libya. They also believed that there were 7,000 British in British Somaliland and 10,000 French in Djibouti; in fact the British had 500 of all ranks from the Somali Camel Corps and a battalion of fewer than 1,500 men of the Northern Rhodesia regiment. The Italians also thought that Britain could reinforce British Somaliland with elements from a 10,000 man garrison in Aden, but Aden had only two Indian battalions. Finally, they believed that there were 39,000 British troops in Kenya – while in fact there were just two brigades of the KAR, and one Indian brigade.

In Kenya there was fear of an Italian expedition marching 400 miles south from Abyssinia towards the Tana river and Nairobi. But from the Whitehall perspective this was as nothing compared with the threat of an Italian invasion of Egypt, which Mussolini was preparing on a massive scale.[14] Furthermore, Kenya was effectively defended by the Turkana Desert, and Churchill argued that the troops which were coming to reinforce Kenya – some 25,000 men from the Union Brigade of South Africa and two brigades of West African troops – would be much better deployed in the Egyptian Delta than sheltered behind the 400 miles of very inhospitable desert. At this stage forces in Kenya were being reinforced. By the end of June two brigades of West Africans had landed in Mombasa. To secure as large a fighting unit at Mersa Matruh in Egypt as possible, Churchill brought the African reinforcements up from Kenya. As he said, the Kenya border was less vulnerable. He wrote of Kenya:

> I had ridden over some of this country at the end of 1907. It is a very fine looking country but without much to eat. The idea of an Italian expedition of fifteen or twenty thousand men, with artillery and modern gear, traversing the four or five hundred miles before they could reach Nairobi seemed ridiculous.[15]

Furthermore, behind the Kenya front lay the broad-gauge Uganda railway. 'We had command of the sea, and could move troops to and from by sea and rail with a facility incomparable to anything that could be achieved by enemy land movements'.[16] So his priorities in East Africa remained Egypt and then Sudan. He was right; the Italians did try to invade Kenya, but only managed to take two British posts in both Kenya and the Sudan – the forts in Mayale Isiola and El Wak, towns on the border of Northern Frontier District reaching about 100 kilometres into Kenya. The desert – arid, treeless and waterless – stretched from the shores of Lake Rudolph in the west to the Indian Ocean in the east and proved an implacable ally to Kenya's small forces.

However, in early August some 25,000 Italian troops invaded British Somaliland from Abyssinia. They were opposed by about 4,000 British troops, drawn from all over Africa.[17] Within two days of the invasion, the Italians had taken Hargeisa and by mid-August the Italians entered Berbera, on the coast, then the capital of British Somaliland.[18] The British were forced to evacuate, carrying the garrison across the Red Sea to Aden, and the British forces withdrew from Somaliland, fortunately with few losses.

By the close of 1940, Kenya's forces numbered 75,000 as reinforcements had arrived from western Africa, Nigeria and the Gold Coast – 10,000 men of whom the vast majority were African; and 10,000 South African soldiers (all of them white). But a counterattack looked difficult since there was no water; the dry land between the British bases and the coastal areas of Italian Somaliland was extremely inhospitable. However, in January 1941 a British geological survey section found water at Hagadera, between Garissa on the border with Kenya and Liboi on the Somaliland border. They drilled down for 350 feet and were able to pump 600 gallons per hour – and enabled the Commander in Chief, Cunningham to launch the counterattack with the battle of El Wak. He advanced rapidly into what is now Somalia, and the 11th Division of Nigerians took Mogadishu by the end of February, a major breakthrough. The ground troops were supported by air cover from the South African Air Force (SAAF) squadron. Troops from Aden regained British Somaliland from the Italians.

The next target was Berbera, where in March Cunningham carried out a successful landing on the beach of two Sikh battalions of the Indian army. Troops pursued the Italians through Abyssinia, and Addis Ababa fell to the KAR in April. At the same time Lieutenant General William Platt launched an offensive from the Sudan, and two Indian divisions marched into Eritrea, squeezing the Italians from two directions. The Italians finally surrendered in May 1941, although Italian forces carried on with guerrilla activities in the north until the fall of Italy in 1943. The Italian leader, Duke Aosta, was taken prisoner in late May.

After the success in Abyssinia, the South African troops went to fight in the key northern Africa campaigns against the Germans, but Kenya was now involved in other theatres. In February 1942 Japan took Singapore, the largest British surrender in British military history. The Japanese followed this up with an Indian Ocean naval raid in March and April 1942 against Allied shipping and bases in the Indian Ocean. This raid forced the British fleet to relocate from Ceylon to Kilindini port. Mombasa's peacetime population of about 1,000 Europeans increased overnight to some 25,000, with naval engineers, storekeepers, dockyard workers, anti-aircraft gunners, Air Force personnel, and all other kinds of military and civilian units. Mombasa was under threat from sea and air alike, and the city took on the air of a fortress, with no one allowed to enter or leave without a permit.

The fleet in the Indian Ocean based off Ceylon was gradually reduced to a convoy escort force, and Mombasa became one of the key bases in the invasion of Madagascar. After the fall of France, Madagascar was under the control of Vichy France, and the Allies were concerned that the Axis powers might use Madagascar as a base – it was strategically placed amid Allied shipping routes in the Indian Ocean. At this stage in the war, Japanese Imperial Navy submarines, which had the extraordinarily long range of 10,000 miles, were moving freely throughout the Indian Ocean. The Allies planned an amphibious assault on Madagascar, known as Operation Ironclad with ships from both South Africa and Mombasa. Following a major assault, Diego Suarez, a port on the northern tip of Madagascar was taken on 7 May 1942; but resistance by the French with Japanese support was strong and low level fighting continued on the island until the capital was finally captured and the island surrendered to the Allies in November. Micky Migdoll, musician and gentleman's outfitter, had reluctantly been among the troops who made this attack.

There is an account of the war years that adds a different slant to the narrative of troop movements and battles won and lost, written by John MacDonald, the local manager of Barclays Bank. His account is rich in details and concerns from a banking business point of view. He is concerned about the management of bank books and accounts, and staffing, since he had lost a good proportion of his banking staff to the forces. Many bank accounts were frozen at the outbreak of war, and a whole layer of regulations and controls on foreign exchange transactions and credit restrictions for importers and wholesalers 'added materially to the work of those bank staff who had not been called up'. As the numbers of troops increased so did the number of account holders and transactions. Many of the troops now arrived from Britain had no idea of how long they would be away nor of where they were going, and so had been unable to arrange for funds to be transferred, as none had East African currency. 'At branches such as Nairobi and Mombasa the staff were occupied for many days after the arrival of new drafts in the opening of new accounts – the record day', he says, 'in Nairobi saw the opening of 83 new military accounts!' He adds 'some departure from normal banking practices was essential and it is pleasing to note that although many cheques were cashed for people of whom little was known, very few came back with the dread reply "refer to drawer"'.[19]

The bank branches also became general information centres – a cross between a citizen's advice bureau and a tourist information office, as newcomers in the forces would ask for advice on where to spend their leave, how and where to see lions and how to obtain shooting licences or films for their cameras.

Once a large number of Americans had arrived, troops brought not only dollars but required 'services which might in the pre-war days have been regarded as not of a banking nature'. Bankers of the old school might, for example, have been startled to read a large notice in the bank worded 'How to send candy and flowers to the U.S.A.'.

Mombasa was also a major intelligence gathering outpost during the war. Between 1942 and 1943, a code breaking unit and listening post codenamed KILINDINI was based at the Allidina Visram High School, overlooking the Indian Ocean. One intelligence officer, Joan Spinks, described how their watch room was constantly plagued by bats and masses of flying insects, and the Arab dhows coming from the sea 'beat drums so loud that we couldn't read signals at times'.[20] There had been Bletchley Park code experts based in Singapore and Ceylon, but they then transferred to Kenya and by November had broken two major Japanese codes for communications between the Japanese Merchant Navy and the Imperial Navy; and by November 1942 the code breakers could read all the messages, allowing the Allies to track and attack enemy traffic.

It was during the war, in August 1941, that Owen sought – and gained – promotion to the post of Chief Grader and Inspector, whose role it was to ensure that all produce leaving the country was clean and of good quality, and equally that imported goods were not carrying pests or diseases. The quality check was essential to maintain the position of the products in international markets. The agricultural officers in the Reserves carried out this function for produce there, but the post at Kilindini port dealt with products from everywhere in Kenya. This was an important administrative post, and brought Owen out of the narrower concerns of the Reserves to a key role in the total agricultural picture. It also brought him from a quiet corner, largely protected from the war, to a city in the hurly burly of all that was going on, and his role was central to what had become Kenya's key function, the provision of food to troops in the Pacific and to the internees and troops in the country.

He initially spent time in Nairobi at the Scott Laboratories for training and familiarisation, and moved with Dorothy to Mombasa at the end of October 1941. The post carried a salary of £690 per year, up from £540 when he was posted to Kisumu, and equivalent to between £30–35,000 in 2010 prices. He loved Mombasa, gloried in the steamy heat, keeping his hands on piles of blotting paper as he sat at his desk so that he could hold his pen and write without it slipping from his grasp.

Mombasa had changed out of all recognition since his arrival at the end of the 1920s. Then, it was a small town with one or two hotels, the Manor (where he had stayed when he first arrived) and the Palace, and very poor communications along the coast. These hotels were still there (with Liza's shell still at the Manor Hotel) but now, 15 years later, Mombasa was well developed as a tourist resort, with the Palm Beach Hotel down the coast at Malindi. And there was still much wildlife – wonderful birds along the beach, curlews, pythons, whitefish and jumping fish, hyenas; fire flies at night. The town boasted two European hairdressers, two movie theatres, the Regal and the Majestic, with two picture shows each week; a laundry and dry cleaner and three petrol stations. It also had a fine cathedral and elegant government buildings on the edge of the Arab old town. Many of these still stand, and Mombasa is taking pride in the preservation of buildings from the colonial period and before; gradually the old town is being restored to something of its former interest and charm. The cathedral dates from early 1900s, and was built with a dome so that it resembles the mosque and other Muslim buildings.

By the following May, once Mombasa had become the naval base for the Indian Ocean and Pacific front, the harbour was crammed with battleships, cruisers and destroyers. Sailors in white ducks strode the city. At one point there were 200 vessels using the port and seven admirals based there. Inevitably there were food shortages from time to time; and all hotels and guest houses were requisitioned to provide accommodation, messes and clubs; Government House (another grandiose Grigg project, financed this time by the Kenya and Uganda Railway Company) became the Officers' Mess.

With the outbreak of war, women were brought into the work force; the Kenya Women's Emergency Organisation (KWEO) was set up under the aegis of the governor's wife, Lady Brooke-Popham. Women, through the KWEO and the EAWL, undertook many activities in support of the war effort, ranging from knitting socks and pullovers, making shirts and clothing; to providing food, sweets and cakes, coffee; to setting up libraries with contributions of books, and starting an entertainments section to put on music and drama for the troops. Many women were left to manage the farms as men were called up, but others, like Dorothy, took on secretarial and administrative roles, where their skills were much in demand. Dorothy herself worked in the Naval Officers' Mess.

Feeding the troops became an important issue. Cattle were collected up in the Northern Frontier District (NFD) or in Somalia – the Somalis were the great cattle traders of the Horn of Africa — and driven down on the hoof to Mombasa to help feed the troops. William Luke, a young recruit to the veterinary service, told of times spent collecting cattle from Mogadishu – presumably after the British had regained British Somaliland – and driving them through the desert, sleeping under the stars beside a thorn tree as he drove his herds.[21]

Kenya was required to supply enough for its now much swollen population – the military, the Italian prisoners of war – and it was also a first line of export to the Pacific theatre. There were nearly 55,000 Italian prisoners of war in Kenya, and in 1942 some 10,000 Poles arrived – now much more than a trickle. It was a time of prodigious effort. The whole machinery of the Department of Agriculture was called on to meet the food needs of the war effort. The British government decided that the products it most needed from East African agriculture were butter, wheat, oilseed, flax, sugar, tea and sisal. The effort meant bringing new virgin lands in the white areas under cultivation – nearly 50 per cent of the land available to Europeans was not being farmed at the time and the government introduced a number of measures to accelerate agricultural production. A production board was set up under the Department of

Agriculture; special short-term loans were made available, and men who had signed up from their farms were extracted from the military if it was felt that their contribution would be greater on the land. This mainly applied to men in their forties. Machinery was supplied from America and Canada. Phosphate fertilisers (without which wheat would not grow in Kenya) had to be found. As there was a shortage, animal bones were collected from around the country and ground up to provide the necessary nutrients. In fact agricultural production doubled during the war, and there was wholesale modernisation and mechanisation in the European areas.[22] A consequence of this, which would have unforeseen consequences, was that many of the squatter farmers and farm labourers were no longer needed and were pushed off the land to merge with a floating and frequently unemployed urban population: the group in society who became increasingly alienated after the war and fostered the growing rebellion which would become Mau Mau.

However, as a consequence of all this investment, by the end of the war the level of sophistication, modernisation and productivity in the European farming areas was extremely high. Immediately prior to the war European farming was still emerging from the Depression and the trial and error period, while Africans were moving hesitantly from a subsistence to a cash and cash crop economy, and 'the two streams were certainly within shouting distance of each other even though voices across the gap were seldom heard. But by the end of the war the two streams had moved so far apart, and the European stream had progressed so much faster, that they were almost out of sight of each other'.[23] This vast post-war difference may explain the seeming invisibility of the efforts in African agriculture that the department and the advisers, not to mention the farmers, had been putting in during the immediate pre-war period.

Once again, compulsory labour became an inflammatory issue. In 1942 the Kenya government passed a Defence Regulation whereby compulsory African labour on European farms was allowed as an essential undertaking for the war effort. Ever direct in his views, Donald Storrs-Fox wrote angrily about the use of forced labour and indeed of the requisitioning of food surpluses where the supply may not in fact have been adequate.[24] His wife Betty reports on his conflict with the provincial commissioner in a letter home:

> D[onald] has been having a war with his provincial commissioner [Fazan] over the question of more or less forced native labour, over which he felt he had to make a stand on principle. I think it has blown over as Archdeacon Owen has got onto it and I think govt has got cold feet over compulsory labour, knowing what a stink it would cause at home.[25]

In the Reserves pressure was also put on the African population to increase output, but this met with resistance from both African farmers and the agricultural advisers. Oates writes about the impact on the soils in the northern area of Nyanza, at the foot of Mount Elgon, a maize growing region.

> The Kitosh were encouraged [by the Agricultural Department] to rest their land, to put it back to permanent pasture and to use boma manure. A model holding was started and various methods of soil improvement experimented with on trial plots. Some success was being achieved when the food shortages of 1942–44 occurred. All attempts at controlling the area under maize had to go by the board and unrestricted cultivation was the order of the day, not only in Kitosh but all over Nyanza. Agricultural officers warned Government of the consequences of this action, but were told that as food could not be imported and as there was danger of acute shortage in many parts of Kenya, it was the duty of all concerned to see that as much food as possible, regardless of the consequences, was grown. As a result, by the season 1944–45 the total number of bags was about 1 million in all Nyanza (compared with 200,000 in 1940). The destruction of fertility caused by this can be imagined![26]

Donald Storrs-Fox was equally incensed about the requisitioning of crops. Betty continues in her letter:

> He [Donald] has also been in disagreement with him [the PC] over the requisitioning of millet supplies, buying up millet crops in this district [Kisii], when there has been a locust warning out, a few months ago, to fulfil an order for Eritrea. He thinks this was an offer undertaken lightly by the provincial commissioner without any investigation into how much to sell and if the people want to sell it. It is the same principle again involved only this time I fear Donald is not going to get the support he had for the labour *shauri*, as I rather think Fazan was involved in accepting the contract without proper investigation, so feels he must produce the goods.... D is standing on a principle and I can't see that God would let this come to nothing over a matter which really means the good and Christian way of doing things for the natives. It is their rights that are involved and when we are fighting for the rights of the individual it is a bit thick to have all the propaganda we do have to tell them about the wicked Nazis, and then do a thing which is Nazi like.[27]

Donald takes up the line:

> [The question] as I see it, is whether we are fighting for the freedom of all people in the world, or whether we are going to chuck freedom overboard for the sake of winning. Freedom is not easy to define, but the aspect on which I have not altogether seen eye to eye with my respected employer [i.e. Provincial Commissioner Fazan] is the free choice of the individual, whatever the colour of his skin, to dispose of his life and property as he pleases within the limits allowed by law'.[28]

In 1943 he joined the KAR and fought for two years. Others who joined and fought were Roger Lambert and Cedric Oates.

Wartime brought considerable privations for Europeans – perhaps shortages of food, but certainly there was acute isolation and a sense of being cut off from home. Communications became more strained, as the dark days of war became darker. For the European population, it was a time of great anxiety, distanced from the dreadful state of affairs in bereft Britain, in contact only by radio. And there was no home leave, of course – people and vessels were needed for the war effort. Owen's father died during the war, a tragic loss and they could not go home to share the grief of their family.

But despite the war, living in Mombasa was good for Dorothy and Owen. Pictures show them on the beach, picnicking, Dorothy standing in the shallows on Nyali beach her arms outstretched, a *kikoi* cloth stretched above her head in the tropical wind; they swam too in the pool at the Mombasa Club, near the harbour and overshadowed by the great stone built Fort Jesus. Their house was a lovely white bungalow with a deep arched verandah, surrounded by frangipani, a higher grade house as befitted his new status. They were driving a Plymouth now, a further step up from the Standard Vanguard of Kisii days. The guests in the Visitors' Book changed however. While some of the familiar agricultural names appeared – Ralph Le Pelley for example, who was the senior research officer at the Scott Laboratories or A. E. Pound who was working in Malindi – two officers from HMS *Royal Sovereign* stayed in January 1942, and Dorothy and Owen played host to a group of officers and WRNS for Christmas 1942.

Owen became very involved with the Mombasa Cathedral choir. The cathedral is a lovely white building, now surrounded by commercial development, but still in grounds with mature trees and a lovely sense of space. In Owen's day there was an old pipe organ which had been 'erected to the glory of God and in memory of the gallant men who laid down their lives in the East Africa Campaign 1914–18'.[29] It is no longer there – one story has it that it was eaten by termites, but another says that it was dismantled and sent to Uganda as the pipes could not hold their pitch in the salty humidity of the coast. A third elides the two stories – termites got to the woodwork, but some of the pipes were sent to Makerere.

Owen was choir master from 1942 to 1945, and he mounted some spectacular music. In August 1943 they performed music from *The Messiah* – 'assisted by many friends among both civilians and the Services'. Perhaps the highlight was a Centenary Festival service in November 1944 to celebrate 100 years since the CMS began its work in Kenya through the indomitable Joseph Krapf. The choir sang music by Purcell ('the Bell anthem'), Stanford, Mendelssohn ('He watching over Israel') and Handel, and Gounod's 'Gallia', a beautiful motet which had been composed during the siege of Paris in 1871 with words from Lamentations, and which Owen and others thought appropriate for a time of war and anxiety.

> Solitary lieth the city, she that was full of people... sorely she weepeth in the darkness, her tears are on her cheeks... Zion's ways do languish, none come to her solemn feasts; is it nothing to all ye that pass by? Behold and see if there be any sorrow that is like unto my sorrow... see how the foe hath magnified himself, Jerusalem O turn thee to the Lord thy God.[30]

The solo was sung by Marjorie Hawkins, a leading Kenya soprano, and Nessie Roper, whose daughter still lives in Kenya – a well-known musician who played organs and harmoniums all over Kenya – played the organ. 'An augmented choir has been assiduously trained by the Cathedral Choirmaster, Mr W. O. Sunman.'

For their 'home leave' in 1944 Dorothy and Owen visited South Africa and Owen kept a diary of the trip. They travelled 9,149 miles, on a circular trip, starting on board the SS *Menaela* to Durban, then on to Pietermaritzberg and Cape Town by rail (South Africa Railways and Harbours) and motor coach. On the return journey they travelled by rail (B&M R) from Cape Town to Victoria Falls and Elisabethville, and then by a series of vessels to reach Kisumu on the SS *Usoga* from Mwanza in Tanganyika, finishing on Lake Victoria. In the first part of the voyage Owen spent most afternoons hand writing out copies of 'Gallia' for the choir.

The boat trip followed the pattern of most voyages; exercises and PT once or twice a day, entertainments and dances in the evening, and bridge, endless rubbers of bridge to midnight and beyond. 'Interesting bridge with 2 small slams bid but not got, in the first two hands', 'bridge late tonight – after being 26 down after 3 rubbers in which we won not a single game we steadily pulled up and through their overbidding, we finished the evening (at 12.20am) only 6 down'... and so on. But there were differences from other journeys. For one thing the vessel sailed in a convoy. They were travelling with a cargo ship – designed for the north Atlantic trade and coal burning, with a European crew – 'the temperature in the stokehold is said to reach 150 degrees!' and escorted by three trawlers keeping station ahead and on the port and starboard bows.

There were about 600 Polish women on board too, aged between 19 and 45. They were to be transhipped to another boat immediately on arrival in Durban and to proceed to England. The diary reads:

> They are all enrolled in the Polish WAAF. Some have come from German-occupied territory, others from early Russian occupied parts. One, a doctor, told how she had been in captivity for 18 months, twelve months chopping wood and 6 months in coal mines. She lost the top of her thumb through a wound going septic. They say the Russian treatment improved somewhat when Germany attacked Russia. They are getting very worried now that the Red Army is advancing into Poland itself. They are indeed between the Devil and the Deep Blue Sea.[31]

Another told of her captivity with her son and daughter, the distress of her railway journey, locked in a truck for 12 hours at a stretch, the lice and the dirt of the Russians.

Some of them entertained in the evening, with a show of choruses: 'they put up a splendid show, unaccompanied, keeping excellent time but being inclined a few times to get sharp. Some very good voices among them'. He notes that they are resilient, but many of the older ones, some of them mothers of the younger girls, look 'so tired and worn by suffering and anxiety'.

Having arrived in Durban, Dorothy and Owen spent some time going 'round and round the shops trying to get our bearings and recover from the bewilderment we have experienced being in a big city instead of a one-horse place like Mombasa'. The shops were full of goods. Some things they wanted to get for themselves or friends in Kenya included football bladders, shut-eye dolls, linen cloths and these were unobtainable in South Africa too, but 'watches, watches, watches by the thousand; chocolates by the stone; and while cotton frocks were thought of as dear at 15/- or so each, that was cheaper than they could be made in Mombasa'. In fact Dorothy bought a watch to replace one that had been stolen the previous year (it had been a twenty-first wedding present) and so did Owen – 'shock, water and magnetic proof Rotary, luminous with steel wristlet – a superb looking piece of work'. They relished music – the South Beach Concert Hall with a full orchestra playing light music; a City Hall celebrity concert – and the movies: Bette Davies in *Little Foxes*; *Suspicion*, directed by Hitchcock was pronounced 'awful tripe', but they saw *Random Harvest* with Robert Colman and Greer Garson, for the second time.

They travelled north overland on the return journey, to Livingstone which was very enjoyable, although arrival in Albertville (now Kalemie) in the Congo was 'the blackest stage of our journey. The Belgians have no idea of travel organisation' – this referring to the Chemins de Fer du Grand Lacs:

> when dinner came it was a curry soup, terrible macaroni, and a tough meat dish badly cooked and badly served… our last Belgian meal was as bad as dinner last night – bread, cheese, uneatable polony and undrinkable tea. For these two meals we were charged the equivalent of 7/- each – Truly our gallant allies know how to charge![32]

They went on to Kigoma in Tanganyika; 'Oh! It was a relief to be back again under a British administration'. The last leg of the trip was on board SS *Usoga*, where they:

> enjoyed a most extraordinary spate of bridge… there were enough players to make up two tables and we played morning, afternoon, evening and after dinner, with the

> Captain, the Chief Engineer, a Captain from the RE [Royal Engineers?], the Chief Justice of Tanganyika, a crown counsel from there, and three of their co travellers![33]

When VJ Day finally came, in August 1945, Mombasa celebrated with a victory parade.

The push on agriculture had immense implications for the land and for farmers. On the one hand the war had ensured good prices for agricultural produce – maize, wheat and beef – and both African and European farmers did well out of the war, so incomes were higher, consumption was stronger. Elizabeth Watkins wrote:

> during the immediate post-war period, for the first time in the history of the colony, there was a sellers' market for agricultural produce. Tremendous improvements had been made in local production, and in raising the quality of African produce so that it met the quality required by canning and preserving institutions. Good prices encouraged production and there was an unexpected sense of prosperity in many parts of the country.[34]

But the price of this success was severe soil erosion and deteriorating soil fertility – in both the European and the African areas. Efforts to reach high production goals caused over-cropping, exacerbating all the problems that had been building up before the war when Denzil Blunt, the Director of Agriculture had written:

> The real interests of agriculture lie in the future as well as the present and efforts are being made to ensure that the methods followed do not achieve present gain at the expense of damage to the soil. *Systems of farming are being worked out which will maintain and even increase fertility and avoid soil erosion.*[35]

But the war intervened and he writes in the annual report of 1945:

> since the last departmental report was printed, six momentous years have passed in which much experience was gained, in which many difficulties have been faced and overcome, in which new organisations have been set up dealing with agricultural problems, but in which *the potential production of our land has deteriorated to an alarming extent.* For some 25 years we have taken from the land all we could get, we have put back in that period negligible quantities of fertilizer and feeding stuffs and the bulk of the land has never had either. The time has passed when it was possible either on native or non-native areas to forsake deteriorated land and break new, and we have to face the issue of building up fertility on our arable lands or cease to exist as an agricultural country.[36]

The intense over-cropping during the war years and the impact of this when combined with two drought years was very serious. The problems were by no means limited to the African areas. Mervyn Hill echoed Blunt's words rather more poetically in an editorial in the *Kenya Weekly News* in 1949:

> The roots of the agrarian problem in the Highlands creep back through the arches of the years to the dawn of time. In the early days very few farmers had enough capital to tackle the immense task of creating a farm from a slice of raw Africa. They had to cheat the land in order to create capital for development; and if they had anxiety about the process it was quashed by the myth of the inexhaustible fertility of Africa's soils, a myth which persisted until it was dispelled by the stark evidence of wide acres ruined by the monoculture of cereal crops.[37]

Governor Mitchell, in his review of the sector[38] later, in 1947, referred to the 'extensive' damage suffered in European areas as a result of 'faulty agricultural and pastoral practices and in some cases... the lethargy, ignorance or greed for money profits of some farmers or... absentee ownership'.

The war marked profound changes and shifts in Kenya, in relations between the government and the African population, and in particular in African political awareness. Something similar had happened in 1918, where Africans fighting alongside white officers as men together found themselves expected to revert to a subservient, inferior place. In 1942 the Labour government in London had challenged the use of compulsory labour on European farms, and this was effective, but, 'remote control could never be sufficient to guarantee African interests'.[39] In Kenya many were coming to see that it was essential there should be better representation of African opinion, and the government proposed a Standing Advisory Committee for LNCs would be appointed. It would have five African members and, it was hoped this would build closer links with the LNCs.[40]

In August 1943 the government felt it needed a missionary 'of a militant type' on Legco (one would have thought Archdeacon Owen had been on the militant side), and Leonard Beecher was appointed to this role. He had been largely responsible for an enquiry into payment of African taxes in 1934 (the Morris Carter Commission), and he articulated his commitment to African rights in an interview with the press on his appointment to Legco: he said he had tried to avoid issuing in any sense a 'manifesto' but he announced that he looked forward to the day when nominated Africans would sit alongside the nominated Europeans and even replace them. As a fluent Kikuyu speaker, he attended LNC meetings and asked Africans to appeal to him by letter – which they did, by the mailbag. He seems to have been very well respected in this role; for example, James Gichuru, later president of the Kenya African Union, had opposed Beecher's appointment saying that 'no European could possibly represent African Opinion' but in 1944 he welcomed Beecher's reappointment to the task.

Other post-war changes brought the appointment in October 1944 of the first African on Legco, the first in East Africa, one E. W. Mathu, the son of a Kikuyu medicine man and a graduate of Balliol College Oxford. Bennett writes:

> Mathu's appointment and the support he had from the political organisation, the Kenya Africa Study Union (later the Kenya African Union) which he helped create in 1944

> marked the emergence of the African onto the full stage of Kenya's politics. He would be courted by Asians and Europeans alike. Whose was the power of the future?[41]

The constitutional and political wrangling would continue. The seeds for change were sown but how would they play out?

6

AFTER THE WAR 1945–50

> 'It is up to us who now hold this land to take urgent steps to ensure that the heritage we bequeath to our successors is not completely valueless.'
>
> AGR, *Director's Report*, 1945

> 'The land is the only thing in the world worth working for, worth fighting for, worth dying for, because it's the only thing that lasts…'
>
> Gerald O'Hara, *Gone With the Wind*

With political changes came a shift in economic policy. The British government had begun to envisage a more active role for itself in the colonies from 1940 with the first Colonial Development and Welfare Act – partly to improve life for the local population, partly to impress the world with our consciousness of our duties as a great colonial power.[1] With the election in 1945 of a Labour government in the UK, the importance of infrastructure and long-term investments in the colonies was recognised, and the 1945 Colonial Development and Welfare Act brought significant funds, bringing to an end the barren rule that every colony was to be self-supporting. The central funds allocated in 1945 were £12 million per year. For the first time, the Colonial Office set out a statement of policy on agriculture in the colonies. Three principals were laid out: the preservation and improvement of the productive power of the country; the social welfare and advancement of the people; and the economic use of the land and labour available, that they are used to produce the maximum return to real value.

Kenya was one of the first colonies to apply for the new funding, which meant that for the first time in its history plans could be made at a colony level with some confidence that funds would be available. Shortage of funds had been a constant lamentation in departmental reports. So, in Kenya, a Development Commission was appointed in 1945, and in 1946, after widespread consultation, it published a ten-year development plan with a total budget of £15.6 million, of which £8.1 million was allocated to crops, livestock, land and water. In 1946, £3 million was allocated to African agriculture by the Labour government – an unheard of budget, a doubling and more of resources.[2] Two boards were set up for the management of agriculture; the Board of Agriculture for the European sector, and the African Land Development Board, known as ALDEV, for the African sector. The structure of ALDEV was quite radical as it cut across boundaries between departments and, as the departmental report recognised, it 'dealt with matters that the department would not normally deal with'. Its plan stressed rural development in African areas and increased areas of cash crops; programmes to build dams and increase water supply, to improve access roads, to clear new settlement areas – these were all supported, and the largest portion of the funding was for soil conservation and the creation of miles and miles of terracing.

For there was much to be done. Huxley writes of 'a condition that was destroying the country's very basis and threatening its future – the bleeding to death of the land. Rivers were running red with the topsoil of the highlands and the stain spread 50 miles out to sea'. Some areas were worse affected than others, of course, and among the very worst was Machakos. Machakos had long been a vulnerable area, prone to famine and drought – this is mentioned in the early writings of pioneers, and the builders of the railway. Since the turn of the century it had early been an important market centre yet, as John Ainsworth had noted, while he was working for the IBEA and establishing a town in Ukamba, 'under the best of conditions, Ukamba's weather weary soil produced a sparse agricultural yield... if the rains came too late the anaemic patches of maize and millet would shrivel'. Drought in 1898 and 1899 meant that Ainsworth was bringing food into the area from neighbouring districts and had to set up relief camps. Starvation prevailed. Mrs Stuart Watt, wife of an English missionary at Machakos in 1900 wrote that 'the scenes round the mission station were appalling – skeletons were tottering hither and thither with every joint and bone in their bodies exposed to view'. By the mid-1940s the same vulnerability was again in evidence. Huxley continues:

> Machakos... was a classic example of all the worst agrarian evils of modern Africa: red, bare hillsides stripped of their vegetation and soil; rivers degenerated into wastes of sand; cattle like walking hat racks; people burning cow dung for lack of firewood; women toiling miles to muddy holes they shared with goats for dribs and drabs of water. The picture in this district was grimmer than in any other because so much of it consists of steep hills easily ravaged by heavy storms... because all the forest cover had gone, even from atop the highest and steepest hills, and because there were three times as many cattle as the land would hold.[3]

But it was a more widespread problem than just Machakos; much of Central Province, large areas of the Nyanza and the Highlands were in very poor condition.

Theories and argument to explain the disastrous state of the soil in Kenya abounded: some blamed bad practice by African farmers; some blamed bad advice by the agricultural officers; some blamed government policies and others the inherent greed of farmers. The accepted wisdom in the settler and government communities was that land erosion was due to population pressure and poor crop rotation, inadequate use of soil conditioners and manure, and over-grazing, the fault of the African. This agricultural team were only too well aware of the problems, but the diagnostics and remedies were less obvious, and some of the agricultural advisers even began to question some of the advice they had been giving.

Thus for example, Nightingale, an agricultural officer who had arrived in Kenya in 1914 put it down to the introduction of crop rotation on modern English lines, whereby a single crop is produced each season, a rotation of maize, cash crop, root crop and legume. Africans traditionally preferred growing a mixture of crops in the same field and Nightingale argued that it was years before agricultural officers realised that the mixing of crops was not only a good means of getting nitrogen into the soil but also shaded the land, allowing soil bacteria to develop and resist soil erosion during the heavy downpours of rain. The old shifting cultivation also stopped a lot of erosion as the abandoned fields covered in bush, grass and weeds held up the rains flowing downhill. 'Our idea', he suggests, 'at least early on, was to plant everything in nice tidy rows which could be kept free of weeds, which we said used up the moisture'. Later, by the 1930s the advice of agricultural officials was, 'congruent with local practices. Crop rotation, composting and grass leys were favoured and proved to be popular',[4] and with the growth of mixed smallholder farming the value of mulch was recognised, such as planting grass between rows of coffee, a technique called thatching. As Nightingale wrote 'Africans were always blamed for soil erosion, but partly because mixed crops were discouraged; all learnt better ways later on'.

Even if practices on the ground were moving in the right direction, there was a strong feeling among the agricultural advisers that they had not been able to do enough, that their needs and warnings had been ignored by a government whose priority was to see more and more output from the soil regardless of the damage this could cause. Colin Maher, for example, a vociferous proselytiser for erosion control and head of the Soil Conservation Service since 1938, put the blame squarely on government policies which encouraged overuse of the land. He had expressed his views loudly for several years. The causes of soil erosion were, he wrote:

> wrong use of land: – wrong methods of farming, wrong densities of population and stock, settlement in the wrong situations. The outward and visible signs of soil erosion are merely the tokens of man in a state of unbalance with his environment. In order to redress the accumulated evils

> which have accrued during 40 years of British rule, with tribal peace and territorial confinement, and to prevent virtual loss of the reserve eventually, for the purposes of human occupation, it will be necessary to spend money on a scale which has not been hitherto contemplated. Attempts to accomplish any noticeable improvement with the meagre funds that have been available up to the present must be as disheartening to the officers in the Reserve as attempting to bail out the ocean with a tea cup.[5]

For all his ability to irritate people – he was described as 'single minded, abrasive and intolerant of manmade obstacles' – he worked tirelessly to rebuild the land.

His perceptions were echoed by Liversage, the economist in the department. He wrote:

> Agricultural officers almost unanimously expressed the opinion that, judged by the results, the British trusteeship for the African had so far showed itself a lamentable, almost criminal failure. The basis of African life was the land. If the productivity of the land was destroyed all must finally perish. Yet we had stood by and let destruction proceed, at an accelerated pace, until widespread hardship and misery was inevitable, not merely at a future time but very soon.[6]

In fact, he concluded this situation had already arrived in certain places and there was 'no remedy except, somehow, somewhere to find other room for the population'. The Agricultural Department report bemoans 'the continued necessity for maximum production of foodstuffs... the agricultural policy has had to remain the same as the Colony was forced to adopt during the war period' because it:

> is having a serious effect on the fertility of the land and already large areas in both non-native and native areas are becoming virtually impoverished and will take many years to rehabilitate. The care and improvement of soil is still

> unfortunately secondary to the continual need for producing crops for surplus and subsistence.[7]

At the heart of the arguments was the fact that developing agriculture in Kenya, whether in settled or African areas was difficult and involved experimentation and risk taking – Delamere and the early white settlers threw tens of thousands of pounds into new ideas – cattle, wheat – and many lost much of their fortune on barren lands and badly understood pests and diseases. In the African Reserves, the early officers struggled too with ignorance – both that of the African peasant farmers and their own as well. The soils looked rich and fertile, but in practice they were vulnerable to the fierce rain storms which could tear deep gulleys into the fields and savagely removed much top soil, particularly where there were large monoculture areas. Monoculture itself drained and impoverished the soils.[8] The land that had looked so fertile and rich to the early settlers turned out to be vulnerable and frail, and both early settlers and technical advisers got it wrong. But whatever combination of factors – poor husbandry, poor understanding of the nature of tropical soils – caused soil erosion, there is no question as to its impact. While this applied to the European areas as much as to the African, the impact on African agriculture was much more devastating; but at last the need for more resources was recognised and funds began to flow.

The belief that most of the African lands were overpopulated and overstocked, and that there could be no major improvement without the large scale movement of people to new lands outside the African areas, had become an article of faith among some of the administration, and it was a widely held belief. Fortunately this idea was not pursued. Hennings wrote, 'it was obvious to travellers by road, by aircraft or on foot that large and fertile parts of the African lands were being hardly used at all'.[9] It seemed to him that there was still good potential in the Reserves if appropriate policies could be introduced and sustained, and this was indeed the way forward.

The possibilities offered by a doubling[10] – or more – in resources would have been very cheering to a generation of agricultural officers who had valiantly tried to make an impact on soil erosion in the Reserves on a shoestring. And there was, by now, a strong team of experienced agricultural advisers and specialists in Kenya, men who had matured over years in the field, changing from raw recruits in their twenties to mature and experienced advisers with a considerable knowledge and understanding of the strengths, weaknesses, and distinctiveness of African agriculture throughout the colony; they were all too aware of the pressures on the land. They included the former soldier settler Bill Lyne-Watt who was the principal agricultural officer in Nyanza and then Nyeri, an inspiring teacher who had worked in cotton, rice, and forestry over his 25 year career; Cedric Oates who had arrived in 1927 at the age of 22 in northern Nyanza Province, and became a regular broadcaster on agriculture and other topics with the BBC General Overseas Service; and Ted Gaddum was still in Mombasa, growing and experimenting with cotton. In 1946 Stuart Gillett was a senior coffee officer and rose to the post of Director of Agriculture in 1948. Michael Grieve was now senior agricultural officer. John Booth had headed Egerton College when it was set up, and later moved back to Nyanza as senior agricultural officer. Tom Hughes Rice was district agricultural officer in Fort Hall in 1945 and went on to be central to the Agricultural Betterment Schemes which were to come with the new resources and new thinking. Owen returned from his posting as Chief Grader and Inspector in Mombasa to join the head office in Nairobi in 1945, becoming Assistant Director and later Acting Deputy Director; he was intimately involved in the planning and development for the agricultural sector which now became possible – something of a contrast to the small and under-resourced plan he had worked on in Nairobi in 1931.

The new financial resources provided the means to attacking the problems, so that there was some cause for optimism among these long serving agricultural officers – Lyne-Watt, Oates, Maher, Hughes Rice and Sunman – that by working together on better resourced policies and strategies they would be able to reverse the soil erosion and rebuild African (and European) agriculture. They were working with a generation of African instructors, who had also developed considerable expertise; some of the more senior and experienced agricultural instructors had been in the department for 20 years, and were considered extremely competent. They were well known, they had considerable prestige in the districts and 'they could by their own personality and reputation do an immense amount of good'.[11]

The new governor, Sir Philip Mitchell, also seemed to be a breath of fresh air, a man of vision. He had a reputation as a brilliant and liberal man, with a strong sense of trusteeship.[12] In his autobiography, he defines the role of the colonial government in the post-war context thus: 'What we have had to do is to help the African to believe in that world, feel at home in it and look with hope and happiness to their future in it'. Mitchell supported the radical reform of the administration concerned with land, with the introduction of district teams, bringing together the Departments of Agriculture, Forestry and Veterinary and Game Services, together with the water division of the PWD, grouped together under the authority of the member for agriculture, animal husbandry and natural resources on Exco and with ministerial responsibility. Agrarian officers were also appointed at provincial level to coordinate agrarian policy in consultation with the provincial commissioner. Medical officers were sometimes involved. Mitchell had become strongly interested in smallholder agriculture, and saw this as the way forward, and at last there was political support for the work with African farmers, and there was room for new thinking.

Tony Swann, DO in Kiambu, writes of the team meetings: 'all the departmental experts were there and could put their varying points of view and when they clashed as they often did you merely talked till you finally got an agreement... we were always able to settle the matter locally'. As well as coordination meetings, the teams used to hold joint safaris and *barazas*:

> seeing sheet tents all in one place made a great impression on the African – you know a veterinary tent, and agricultural tent and an administration tent. 'Oh, they're all in our area are they – interesting!' and they knew you [sic] couldn't

> play one off against another and say that the veterinary officer told me the other day I must have grass and not coffee – ah you said, well you remember seeing him here last Tuesday and they both agreed that it would be coffee and not grass, or vice versa.[13]

This level of cooperation within the government sector was new and proved effective – Tom Hughes Rice said openly at this period (when he was at Fort Hall) that he felt it was the first time he had ever really had the wholehearted backing of the administration.

The central planks of the new policies were to continue with destocking, to encourage terracing and the use of mulch and manure, and the consolidation of land into individual plots. Compulsory destocking had been tried with some success among the Masai, but in other areas, notably Kiambu, met with total hostility. The Kamba, the tribe at Machakos where the cattle population was many times greater than the sustainable level, refused to part with even 'the boniest of the walking hat racks' and marched to Nairobi in large numbers, men, women and children, where they settled down in the grounds of Government House. Eventually they were moved to the racecourse, but further than that they refused to go and the authorities gave in – compulsory destocking was dropped.

Terracing and mulching were more successful. In his rich memoirs Hughes Rice describes how he set up a Communal Soil Erosion gang to work on 'some spectacular soil erosion horrors' in the Sagana River valley when he was posted to Nyeri after the war. He was again working with his friend and mentor, Bill Lyne-Watt, who was now provincial agricultural officer. Tom was always something of a showman and creative in his teaching methods. Working in Fort Hall he collaborated with his DO, Noel Hardy, an ex-Ghurkha officer and excellent artist. He and Noel made a convincing 'double act' and he writes of one particular 'turn' which always went down well – on the use of manure. They worked with a blackboard and illustrations, so Tom would ask Noel to draw a cow, which he did with a few sweeping strokes. Then he would ask what the first big value of the cow was – and Noel would draw a figure milking into a bucket. Then he would say, what about the second value? And Noel would alter the cow's tail to the erect position and out would come the plop-plop-plop of the cow pat. This normally raised a laugh, and, together with the Agricultural Information Department, Tom could raise enthusiasm for the needed communal labour – no more than three hours a day and for two days only so as not to build any resentment by interfering with people's day to day lives too much – with men with ploughs and oxen breaking the terrace, and women making the terraced banks with the loosened soil. After two years the district was quite transformed with most of the eroded hills neatly terraced.

Terracing and manuring in the native areas was making considerable progress – by 1945 there were 6,922 miles of narrow base terraces, built mainly by voluntary labour and community organisations (a small proportion paid for through the LNCs) – and in Kiambu the agricultural officer notes that:

> there has been a rapid awakening taking place to the urgency of maintaining soil fertility, and already ample proof has caused the tribesmen to purchase £2000 of imported *boma* manure... and in upper areas more and more owners are housing their beasts at night for the improvement of their health and for the manure.[14]

Land consolidation, the other important plank of agricultural policy, was also beginning in the late 1940s, and was increasingly understood by the farmers as offering advantages. For example, in 1948 there was an incidence of tick borne disease in Central Province, which spread rapidly because of the communal grazing of cattle.[15] But the agricultural advisers pointed out that if the herds were isolated from their neighbours by the enclosure of holdings then the spread of the disease would be reduced. Furthermore, enclosure helps in the collection and use of manure. One chief was convinced by the argument, and persuaded his people to exchange their scattered fragments among themselves to achieve compact holdings which could be enclosed by fence or hedge. This move was of course supported by the agricultural officers, and led to development of more planned production and management of holdings. Coffee and pyrethrum were being planted at the end of the 1940s, both of which needed to be planted on bench terraces and well manured; these new crops in African areas encouraged both the use of manure and soil conservation measures.

Thus the five years after the war saw the beginnings of a major change in the pattern for African agriculture. With the 1945 infusion of funds, development both at a practical and a theoretical level could be implemented and the new resources and renewed efforts did manage to transform much of the land. Land consolidation, enclosure of cattle, use of manure, terracing; this armoury of tools began to rebuild the land. There was a particularly convincing project to restore eroded land in Machakos, and it turned out to be exemplary, perhaps surprisingly given the awful devastation that has been described in the area. A project known as 'the Makavete Square Mile' in Machakos district demonstrated how eroded land could be restored in the space of two years (which apparently were drought years) and raised the estimated carrying capacity from one beast to 30 acres to one beast to 2 acres. This was a project which was started in 1949, and an African 'betterment team' went in to change the way everything was done. Agricultural improvement was still viewed with suspicion, but the team leader here was one John Malinda, a district assistant, an effective leader who managed to convince his people in the area that this was 'an opportunity for them, not a sly official trick to steal their land'.[16] By 1951 the grass was waist high and the Makavete Square Mile was fully terraced.[17]

Building personal confidence and trust was always seen as key by the *bwana al shambas*, as the agricultural officers were known, but the 'first grumblings of Mau Mau' were becoming louder in the Kikuyu areas. Although more cash crops were introduced by the department – tea, coffee, pyrethrum and vegetables – and this was a period of considerable transformation and recovery of African agriculture, in the Kikuyu districts, Mau Mau activities were, in Tom's words, 'boiling up'. Tom still went on traditional tent safaris but was now accompanied by a police bodyguard – 'carrying an enormous revolver and with size 12 boots'. Much of the work the department was doing with its African colleagues was welcomed or at least accepted, but during these years – the late 1940s and into the early 1950s – the Mau Mau movement was growing, although even at its height it was broadly limited to Kikuyu areas, and many writers note the normality of life west of the Rift.[18]

At the heart of the underlying political issues there was always land. Land is the underlying issue which defines Kenyan politics during the colonial period, and over 50 years had been the principal focus of mistrust and disagreement. From the earliest days, when the administration had alienated 60,000 acres of Kikuyu land in Ukamba province at Kiambu and Dagoretti for settlement; to the forced moving of Masai from their traditional grazing grounds; to the acquisition and development and fierce protection of land by the settlers; to the failure of the government to secure land for native farmers under the Native Land Trust Ordnance 15 years earlier: the most potent issue was always land. The British government had tried to make amends and compensate, particularly through a major review of land chaired by Sir Morris Carter in 1934 – the Carter Commission – but the recommendations were, in reality, too little, too late. The Kikuyu felt they bore the weight of these concerns, and they bore it with growing resentment.

By the 1940s there was a considerable body of educated and aspirational Africans. While education was far from universal, by 1948 there would have been some 260,000 pupils in primary education,[19] and nearly 6,000 in secondary. Mission schools had long been at the heart of education in Kenya, education being a central component of the missionary objectives, although they were of variable quality. Almost all the primary school places were in missions – Catholic, Methodist, Church of Scotland, CMS – while there were only 1,500 children in government primary schools. The first secondary school for African boys, Alliance High School in Kikuyu, was opened in 1926, followed by Maseno and Yala in the late 1930s and the Catholic High

School in Mangu. By 1945 there were two secondary schools for girls. The elite secondary schools were able to feed some graduates to college; Makerere University, first established in 1922 as a technical school for the three colonies of East Africa, became a university affiliated to University College, London in 1949.

Some of the African leaders were educated at British universities, but this did not earn them respect with the colonial government who laid up trouble for itself.[20] For example, Peter Koinange, founder of the Kenya Teacher's College, had been at Cambridge for a year, followed by Harvard and Columbia Universities. He returned to Kenya, full of great expectations, only to be offered a teacher's job at half the pay that its former occupant, a European, had been getting. Incensed, he refused to accept the post, and the government Education Department refused to move; the upshot was that he started the Kenya Teacher's College himself. More fuel for resentment. Other British educated Kenyans included Eliud Mathu, the first African on Legco, who studied at Balliol College, Oxford; and Jomo Kenyatta who originally went to England in 1929 as a paid emissary of the Kikuyu people to present a petition to Parliament against an attempt to impose forced labour on the Kikuyu (in fact, as discussed elsewhere, this was never imposed thanks to the efforts of Archdeacon Owen among others), and then studied at SOAS (the School of African and Oriental Studies) in London and then went to Russia, and was outside Kenya for 17 years.

There were a number of grounds for resentment, some historic – the alienation of Kikuyu land, and later the bitter resentment of the Christian missionaries who attempted to ban female circumcision in the crisis of 1928–31. More contemporaneously, wages on white settler farms remained low after the war, and increasing population accelerated the dismal spiral of land exploitation and soil erosion, pushing more people off the land and into the harsh climate of Nairobi where the predominantly Kikuyu labour force was squeezed by rapid inflation in the prices of basic commodities and squalid living conditions, along with a mass of floating unemployed drifting in from the Reserves and settled districts as labour was pushed off the estates. Indeed, there was a strong understanding in the administration that the Mau Mau uprising had begun among landless Kikuyu working in the Rift Valley and spread covertly to Nairobi and adjoining districts.[21] The Kikuyu, particularly the landless ones, felt alienated from the colonial mission, even as development and African welfare was becoming more central to Whitehall aims. For example, there was African criticism of the 1946 Report of the Development Commission; Leonard Beecher was the Member of the Commissions representing African Affairs from 1943 to 1947 and he was highly critical of the report. 'Nowhere', he said, 'is equality of opportunity in that report applied to African people'. 'Again and again', he went on, 'the recurring phrase "the utilisation of African manpower" as if they were so many checkers on a board. I hate that phrase and I know my African friends hate it even more'.[22] With his fluency in Swahili and Kikuyu he was much respected and 'represented African interests at a time when few other Legislative Council members had any social contact at all with Africans'.

Since the circumcision crisis, many Kikuyu had set up their own schools separate from the mainstream mission schools, and these, in turn, became an arena for Kikuyu debate, discussion, politicisation and perhaps militarisation. The main Mau Mau crisis – the Emergency, the oathing ceremonies in the forests, the burning of villages and the Lari massacre, and the torture and abuse of Kikuyu detainees in the British camps – these were still in the future, but by the mid-1940s there was visibly growing unrest, and the leaders were directly challenging the agricultural reforms the department was seeking to make in the Kikuyu areas.

From the perspective of the agricultural officers, there appeared to be good progress; by the late 1940s terraces were being built throughout the country, usually with communal local labour, enclosure and enclosed pasturage of cattle (to enable the use of the manure and the controlled grazing of the animals) and other soil conservation measures were being put in place, to an extent that seemed well accepted – one departmental report says that different locations were competing with each other to see which villages could terrace the most, fastest. But suddenly, in 1948, parties of Kikuyu women who had been working in the terraces downed tools and refused to work anymore. Men who had requested cattle dips refused to use them, saying they poisoned the cattle. Kenyatta had persuaded the workers that the measures to control soil erosion were colonialist tricks designed to steal the land; and he set many Kikuyu against them.

But although there was clearly an awareness of growing unrest in Kenya, particularly in the Kikuyu areas, the response by the administration was curiously muted. Tony Swann, DC in Nyeri, saw these actions not as political but simply as part of the traditional and pervasive fear that good was done to the land only in order to make it useable by whites, particularly true in the districts near the white settled areas. He expressed it thus:

> if you did anything like putting in a dam or even, when we started cash crops, tea and coffee, you are doing this so that you can give the land to European farmers and that's when your political opposition came in, and that's where you could swear black, blue, we believe it from you, you may be sincere, but once you go it'll be gone and this was the reverse side of the penny, this terrible fear that nobody did anything for nothing and if you did it [the land] was going to be taken.[23]

But the traditional fear was changing to a more articulated political need.

Lonsdale[24] suggests that it was the very training of the district administrators which prevented them from understanding the potential significance of the Kikuyu actions, the resistance, and the underlying political demands. Colonial administrators had no room in their vision of their role as trustees, which entailed a moral obligation to use their power on behalf of the common interest, for the disruption of social conflict or 'the disease of politics'.[25] Administrators were also trained to be dispassionate upholders of law in the context of local affairs and have little interest in politics – and they were still drawn mainly from the ranks of classicists. Indeed, civil servants in general would prefer to carry out their work in peace without the apparently capricious and populist whims of politicians. There was also a feeling in the administration that they had positively encouraged local African participation in local affairs, planning and policy, through the LNCs, which had been in place since 1925 and were an important level of local government and administration; in agriculture, education and other local services. But in political terms, the LNCs could be seen as appointees of chiefs and elders who would simply endorse the wishes of the district commissioner. Such a body would not satisfy the political aspirations of the growing articulate and educated class of Africans, particularly the Kikuyu, always the first tribe to take on change and to recognise the value of what the Europeans had – riches – and that it could be theirs through a mix of education and technology, plus political space.

By the mid-1940s the pressure for a proper, recognised political voice among the Kikuyu was strong. Emerging originally from the female circumcision crisis, the KCA challenged the more passive role of the LNCs and, linked to the Kikuyu schooling system, was seen as potentially trouble making but not taken seriously as a political movement. Berman[26] suggests the activities of the KCA were viewed as 'irresponsible agitation intended to dupe the ignorant rural Africans into believing that the association represented their true interests'. The association of the KCA with the issue of female circumcision was seen by the administration as evidence that the organisation was a reactionary group, hostile to change, modernisation, and development. Hidden in this attitude was a conviction that 'various religious sects and groupings were manipulated by the African politicians', and that there was a strong connection between religious cults and political protest; the administrators tended to see the defence of tradition as a total rejection of progressive change. The converse of this attitude was that demands for change through the KCA were largely rejected by the administrators. But the rapid social changes and economic growth, as well as the extremely unbalanced distribution of the fruits of economic development, increased the intensity of grievances felt and the need for political articulation. The Kenya Africa Union, founded in 1944 was concerned with agricultural development, credit facilities, labour conditions and wages. In 1945 Jomo Kenyatta wrote:

> What we do demand is a fundamental change in the present political, economic and social relationship between European and African... the Africans make their claim for justice now, in order that a bloodier and more destructive justice may not be inevitable in time to come.[27]

Perhaps the combination of the strength of the white settlers at the heart of the Kenya government as well as the traditional views of the district administrators made it impossible for the colony to make the political changes that were needed to ensure the effective development and welfare of the colony. The demands for political change were increasingly articulate, but the Kenyan government seemed incapable of response, and at the heart of the administration Sir Philip Mitchell, despite his reputation for progressive thinking, seemed to ignore officially what was clearly seen by many.

For the administrators on the ground, particularly those who travelled around their areas on a very regular basis (and the agricultural officers and their African colleagues the instructors) were very much aware of the beginnings of a Kikuyu uprising. The missions would have known too, the Christian Kikuyu being caught between the competing faiths of the missions and the more militant Kikuyu. Beecher with his fluent Kikuyu and many of the church leaders on the ground knew that there were changes and trouble afoot. But there was a well-documented reluctance at the centre of the government to respond to the growing movement and a sense that – in the words of Sir Philip Mitchell himself:

> it is simply a question of 'the political agitation of the few' which can be best answered by steady improvement of the standard of living in those areas... firm enforcement of the law and full government support of established authority,

> coupled with adequate security precautions and a refusal to be 'rattled'.[28]

In other words, business as usual.

Even as the evidence of trouble on the ground became clearer, Mitchell continued to turn a blind eye. Indeed it is suggested that part of the problem – from the British point of view – was that by the late 1940s Mitchell had lost some of his incisive brilliance and vision; he had been working solidly for seven years, having been persuaded to stay on for two extra years after his five-year tour, and after a period of nervous exhaustion was apparently drinking heavily.[29] More than one commentator has argued that the administration knew all too well what was going on, despite official denials, and gossips suggest that one of the reasons for the denial was that Mitchell did not want to end his governorship with a crisis; furthermore that he did not want to discourage the visit of the then Princess Elizabeth and the Duke of Edinburgh to Kenya in 1952. The most famous of their destinations during their stay was the Treetops Hotel in Nyeri (where 'she went up a princess and came down a queen'). That part of the story is well known – but less well understood is that Treetops is in the heart of Kikuyu country, so to send the couple there should have been seen at the very least as a serious risk. Tony Swann was DC at Nyeri at the time and wrote:

> what we went through – whether the Queen [sic] and Prince Philip came up at all [to Nyeri] – whether they flew up whether they motored up – I mean it was endless. You spent your nights on the telephone – people coming up from Nairobi going over the route, seeing where you placed people… we had Special Branch and the commissioner up – they all came and looked at the route, and we looked at the Royal Lodge and we looked at Tree Tops and they said 'well it's too late now – you can't cancel now', and by the mercy of God we got away with it – but I was in a cold sweat.[30]

Those with a more malicious tongue suggest that Sir Philip was in denial partly because he wanted to retire with the Baronetcy which tends to follow a successful royal visit. More generously, Huxley suggests that he had been a brilliant governor, fluent in Swahili and far sighted about the potential of Africans, but he:

> had returned from the wars an exhausted man and... was on the verge of a nervous breakdown.... The cutting edge of his intellect had been blunted and that was why when tremors of the impending Mau Mau crisis began to shake the country he insulated himself from their shock, brushing aside the warnings he received from his subordinates and why he ended his career in 1952 under a cloud instead of the bright sunshine that his former achievements deserved'.[31]

Mitchell certainly never went to Nyeri Province to take a look.[32] Perhaps he was unable to face the thought that his confidence in the development of the African did not bring them the betterment and wellbeing that he had anticipated, perhaps he felt some sense of betrayal.

The upshot was that Sir Evelyn Baring was appointed as Governor in 1953 and took very firm steps – some say too firm – but at least based on a much better understanding of what was going on. He visited every district in Kikuyu within a fortnight of arrival, and sought advice widely from experts such as Beecher to explain what was going on. He valued the knowledge of those who had been on the ground for some time – for example, there was a British teacher who had retired after 20 years in Kikuyu and Baring asked him to assess what proportion of the local population was implicated as 'we don't understand what is going on'. The question was between Christian and non-Christian Kikuyu; was there a straight divide between the Christian Kikuyu and the non-Christian Mau Mau? The teacher estimated at the time that, although the majority of the population was actively Christian, only 200 or 300 people were untouched by the movement – the rest were sitting on the fence and the Christians were very torn.

Yet against this background of growing political tension and the poverty caused by land erosion, life in Nairobi for Europeans continued in its pleasant and social vitality. Nairobi had grown out of all recognition, the population more than doubling from about 50,000 in the mid-1930s to about 110,950 by 1945, of whom about one tenth were whites and one quarter Indians. The city was lively and more mature than 15 years earlier with many new services and facilities. European and Indian cultural life was thriving in Nairobi. By now the European population of Nairobi was large enough to support three English language newspapers: the *East African Standard*, the *Sunday Post* and Mervyn Hill's *Kenya Weekly News*. Edward Rodwell, a sensitive journalist in Mombasa wrote a delightful column called 'Coast Causerie' which documented the nature of Mombasa, its climate, peoples and animals. There were more restaurants – Chez Gaby, Falstaff, Jack Frost, The Lobster Pot or the Horseshoe and the hotels – New Stanley, Torr's, Norfolk – continued to thrive. Hairdressers – Pompadour – and dress shops abounded; Micky Migdoll's Esquire continued to dress the most elegant of men. The government servants patronised the Nairobi Club, set up on the hill near Government House and the cathedral. In 1946 the Nairobi Game Reserve was created, 40 square miles, off the Langata Road and beside Wilson airport.

It was also, perhaps surprisingly, a golden period for classical music, for the war brought a number of unexpected musical benefits. Many of the Italian internees were extremely musical and artistic; they had made their instruments while in the camps from whatever materials they could get hold of, and put on concerts in Nairobi. They made their debut as the Central Italian Orchestra in 1943, under the baton of Giuseppe Gagliano who was a great musician and 'as a conductor he portrayed a commanding figure on the podium; he had long black hair and when he bowed, it flopped down below his head, but when he went upright again, it fell back into place as if it had been manicured'.[33] Owen went to one of their concerts in Nairobi in May 1944 and he writes: 'went to an Italian PoW concert – good on the whole, although some imbalance between wind and strings. But the quality of the tone and precision were very good, the solo cellist being outstanding'. The *East African Standard* wrote: 'Many of Gagliano's players are not professional musicians but they provide an excellent example of what can be achieved by musicianship, orchestral discipline and competent leadership in a combination playing regularly together'. In fact, after Gagliano returned to Italy he became chief conductor for the Rome Radio Orchestra.

As well as the Italians, the war brought Bronslaw Fryling, a refugee from Poland and a violinist of some quality, and Ivor Keys, a talented English musician and piano teacher, a Fellow of the Royal College of Music and accomplished organist who was posted to Kenya in the Pioneer Corps.[34] As practising Anglicans, Keys and Owen would have known each other and sung and played together at All Saints Cathedral. Keys later became very well known as a musician, composer, author and broadcaster, and maintained an interest in music in Africa among his other musical achievements after he had returned to the UK in 1947. Fryling played the violin with the Central Italian Orchestra. Another musical talent was Nathaniel (Nat) Kofsky, also a violinist and a fine musician. He arrived in Kenya with the military, and was appointed musical director of the Red Cross. In this position, he travelled around the country with Ivor Keys, entertaining troops and civilians alike and raising funds for the Red Cross. It was a musically fortunate coincidence of people. In 1944 Kofsky, Fryling and Billy Isherwood (the founder of the Nairobi Philharmonia Society in the 1930s) established a Kenya Musical Conservatoire, which was set up in three disused army huts on Kirk (now Nyerere) Road. Fryling became the first director of the conservatoire. Kofsky went to South Africa to lecture in music at the University of Cape Town, but when he returned in 1951 to conduct the Nairobi Orchestra for a single concert he was offered the position of Director of the Kenya Conservatoire of Music, a post he held for 33 years. Although there were not initially many pupils, the conservatoire has survived and indeed thrived, and from the first its doors were open to people of all races – an 'adventurous and farsighted policy' in 1944, but one which Hyslop would write 20 years later 'is now rightly taken for granted'.

By now Owen's great passion and, in musical terms at least, greatest contribution to Kenya lay in the life of All Saints Cathedral and the choir which he built up more or less from scratch as Master of the Choristers. His photo still hangs in the vestry of the cathedral, and for this he is still remembered. He had sung in the choir in 1933 but now returned from Mombasa to lead the choir. He formalised the choir in many ways, one of which was the introduction of choir robes in 1947. Dorothy, who had kept a diary for this year, notes with glee that 'we wore our robes for the first time at practice, and then started wearing them in services on 26th January'.

The first foundation stone for the cathedral – the Cathedral of the Highlands – had been laid in February 1917, and the church built to the design of architect Mr Temple Moore FRIBA, then said to be one of the most outstanding architects in the Gothic style in the late Victorian era. The Englishness of the design and the philosophy is so strong.

> The first portion [of the cathedral] owed its existence to the devotion and self-sacrifice of the few English residents in Nairobi at the time. They were unanimously desirous that the traditions of religious life 'in the home country' should be maintained in this new country, six thousand miles away.[35]

But in the first tranche they had only enough money to build the nave and the chancel arch. The Provost, Hugh Evan Hopkins describes his first sight of it when he arrived in 1947.

> I had got a bit of a shock when I first saw the cathedral which I was going to be in charge of for the next eight years. It really was an ugly sight, only half built, with an unsightly blank east wall where the building abruptly ended without chancel, choir or vestry.[36]

By the time Owen was Master of the Choristers there were ambitious plans for the completion of the building. He sat on the

Cathedral Church Council and the Cathedral Building Committee, and put in a strong bid for a song school to be incorporated in the new designs – big enough to incorporate 'replicas of the choir stalls, and to be fairly lofty, since the singing conditions in the present Old Deanery are frankly appalling – it is well-nigh impossible to get good tone there and the room rapidly gets stuffy'. To this end he wrote an eloquent submission, saying that the musical side was essential to the proper conduct of the cathedral. Sadly, his proposal received short shrift from the chairman, the Provost Hugh Evan Hopkins. 'The matter was discussed' note the minutes, and the chairman emphasised that he felt that an additional song school project could not now be included in the proposed extensions; he considered it would be a luxury. 'The sum of £50,000, it was hoped, would see the completion of the whole building but this was uncertain and the demand for funds could not be increased.' The minutes go on to say that the meeting closed with prayer – but it was too late for Owen's project.

The extension to the cathedral was designed by architect Richard Hughes, at the request of Leonard Beecher, who was now assistant to the Bishop of Mombasa, and also serving on Exco for the government (from 1947 to 1952). Richard Hughes was the son-in-law of Mervyn Hill, writer, chronicler of the railway, manager of the stud in Kenya and – earlier at least – wild child. While Hughes was studying at the Architectural Association in London for his qualifications, he designed a multiracial town and low cost housing with Kenya in mind. He and Leonard Beecher were old friends in Kenya, and when Beecher came to know about this architectural vision, he asked Richard to help design the new cathedral chancel. One thing led to another, and over the next decade Richard Hughes designed 16 or 17 Anglican churches all over Kenya: at Kilifi, Kisumu, Nairobi, Meru, Embu and Nanyuki. He designed a chapel for the Alliance Boys School, the African school in Kikuyu, for £2,000, which is regularly acclaimed as an outstanding piece of architecture. He gave up this part of his practice in the end because everyone wanted him to work on a *pro bono* or at least heavily discounted basis. In later times, in the early 1960s, Richard designed a good many large projects in the downtown area of Nairobi, as well as a beautiful family house on Polpone Avenue. The ceremony of the laying of the foundation stone for the cathedral extension took place in 1949, just before Owen left to retire to England, with glorious music and a procession.

The choir achieved considerable fame and recognition; when the Duke and Duchess of Gloucester visited in 1950, to present Nairobi with its charter as a city, there was a magnificent service in the cathedral at which the 'famous choir excelled themselves as they have always been able to do'.[37] Sunday services were regularly broadcast on the radio – a friend of Evan Hopkins who was at the time in Hargeisa (Somaliland) recalls hearing the broadcast and being much moved by both sermon and music.

Hugh Evan Hopkins arrived as Provost from South India where he had spent five years. After this experience, he was shocked by the extent of a colour bar effectively operating in Kenya. As he writes in his memoirs, the cathedral was the headquarters of the white Christian community, with all the services in English. In those days most Africans worshipped in their own languages at the CMS or other mission churches. He says:

> those who have seen Nairobi or its cathedral in the past twenty years [he is writing in 2001] would never believe how things were in those days when, for instance, on making friends with a Kikuyu who had got a good degree at Balliol, Oxford, there was not a hotel or coffee shop into which I could invite him to share a cuppa! In the 1940s and 1950s the colour bar (about which I preached from time to time, and which did not bring me much popularity) was everywhere. The missionaries worked among 'the natives', while chaplains (like me) worked among the settlers and Europeans generally.[38]

Sitting on the Cathedral Church Council, Owen played a strong part in the management of the Church. Again this was a time when he and Dorothy made lasting friendships (Dorothy sang soprano in the choir). With Beecher and others he founded the Kenya Church Music Society which he chaired until he retired in 1950. One of his great friends was Graham Hyslop who was fascinated by traditional African music, and sought to blend the European four part choral tradition with African melodies. He later played an important role in encouraging African musicians to learn music in the Western classical tradition, men such as John Kimale, Philip Angore, the organist at Mombasa Cathedral, and Washington Omondi, now head of music at Kenyatta University. These men are now custodians of that tradition.

Hyslop had arrived in Kenya in 1936 as an accountant to the CMS and during the war he performed in the entertainment corps in the Middle East. In his own words, in 1944 he formed an Education Corps Unit with responsibility to provide music and dramatics for the East African troops serving throughout the whole of the Middle East – from Mersah Matruh in the Western Desert, all down the Suez Canal and Delta region and up into Palestine and Trans Jordan. In 1953 he returned to Kenya and began working in adult education, particularly the training of junior community development staff, at the old Jeanes School – and put on vacation courses in music at Kabete.[39] He played the organ at All Saints and also taught music at the Alliance School, and played a prominent role in the musical life of Kenya. He later joined the Ministry of Education as inspector of music and drama, but his main claim to fame today in Kenya is that he chaired a five person commission to write a national anthem for Kenya before Independence, at the request of Jomo Kenyatta. Washington Omundi and John Kimale were members of the committee; the anthem is based on a lullaby from the Pokona tribe, who live by the Tana River. (At the time the details of the composition were kept secret – it was deemed far too inflammatory among some of the settlers and other colonists that a white man should be collaborating with Kenyatta for a national, post-British, anthem.) But in the 1940s and 1950s he was an inspiring musician and teacher and was influential in trying to make good music – both African and Western – strong in Kenya.

As well as developing a richer cultural life, colonial officers were now, after the war, able to take home leave again. (The Director of Agriculture, Denzil Blunt, notes somewhat sourly that 'the resumption of overseas leave, of which indeed the whole staff was greatly in need, coupled with the absence of any recruiting during the war, has put a heavy strain on the Department in trying to keep its various services in operation.')[40] Owen and Dorothy set off for the UK in February 1946. But shipping was still tied up in the war effort – the war in the Far East would not finally be over until April 1946, even though the Japanese had surrendered in 1945 – so they journeyed down the Nile from Uganda to Port Said and thence home. Owen kept a diary of this trip too, and they actually kept it (my job would be much easier if they had not destroyed so many of their records!).

It was a journey that took them nearly six weeks, starting by train from Nairobi to Jinja, where they left the Kampala line and went north to Namasagali on Lake Kyoga, to pick up the SW *Stanley*, an oil burning river paddle steamer. Another comment in the diary was tossed out at the Belgians: 'the Stanley is in fact like a pocket edition of the Congo steamers, but being British, very much better, of course'. The *Stanley* had to push some half dozen barges or lighters, lashed three abreast, laden with sugar, petrol, some Indian and African passengers and a deal of livestock, along the line of the Nile through Lake Kyoga – because of shortages of fuel and boats. They made slow progress. There were rubbers of bridge, the occasional hippo and crocodile; as they chugged towards Pakwach 'everything was just delightful with the sparkling water, bright sun and cool breeze. Sitting for'ard on deck chairs, lazing; what more could one want! .... Elephant. Lots of them. And then insects! Lakeflies by the million and mosquitoes by the thousand! Also ants!!' On to Juba ('the biggest place we have seen since we left Kenya but that is not saying much') by car and truck.

Once they reached the Nile the water was so low at that time of year that progress was slow. But the scenery charmed ('round Mogdalla the river banks are really lovely... covered with long grass and many trees – mangos and others, looking not unlike the upper Thames'), bridge was played ('played all evening in the enclosed deck for'ard of the dining saloon. Rather late to bed'). Then through Dinka country ('the Dinka being tall and slender and wearing little in the way of clothing – with immense herds of cattle which go in for enormous horns') until they got stuck on a sand bank; bridge before lunch, before dinner and after dinner.... After two weeks they finally reached Khartoum, a much more dusty and Sudanese landscape now with mud brick square houses and the occasional camel. The Grand Hotel in Khartoum seemed not to live up to its name, with poor management and rooms not ready; and the town itself was a bit of a disappointment with no large shops, just *dukas*.

The trip was very sociable and merry – they were travelling with Robin Wainwright and his wife, whom they had known since Nyanza days when Robin was DO. The hotel turned out to be a very cosmopolitan place with Greeks, French, Egyptians and Sudanese, and a former colleague from Kenya who was Director of Agriculture in Uganda. Owen was surprised by some of the differences between Kenya and the Sudan, noting when invited to a tea party: 'To the March's [government acquaintances] for tea at 5pm; shown to our great surprise into a lovely garden where a 'party' tea was laid out and 2 guests with 3 children assembled. A very nice tea but we were shaken to the core by the way everything, even to the pouring out of the tea, was done by the servants! The Sudan people strike us as very smugly superior – how the women occupy their time is beyond our comprehension!'

On by train through the desert to Wadi Halfa – 'vivid green irrigation plots, palm trees and quite a big and attractive looking place'; then steamer to Aswan and Luxor, and thence on to Cairo, where they stayed on a houseboat with a cabin just on the water line. They immersed themselves in Cairo life for a couple of weeks – theatre (*It Depends What You Mean*, *School for Scandal* and *The Barretts of Wimpole Street* with Roger Livesey and Ursula Jeans, all presented through the British Council[41]), antiquities, cathedral, shopping and walking – until 21 March when they took the train to Port Said (via Ismailia and the Suez

Canal) and then the SS *Almanzora*, an old troop ship, to Toulon. Through the Mediterranean they took a rather long route through a swept channel along the coast to avoid mines. From Marseille, on an American leave train ('the sanitary arrangements are just terrible – no water in the taps but plenty on the floor and the smell!!!') to Paris where huge shortages were evident – of milk, meat, chickens, coffee and even wine; 'the French becoming by force of circumstance beer and water drinkers!' and where they were put up in the 'wretchedest hotel in the Latin quarter, a 15th rate inn'. Then a train to Dieppe, breakfast (bacon and egg) and crossing to Newhaven: 'oh! The joy of being in England again'.

It was very different from the England they had left in 1938, their previous home leave. The ravages of bombing, austerity and rationing must have shocked them. There was the additional pain at the death of Owen's father. Communications during the war were infrequent, letters taking weeks to arrive; and people in Kenya spoke of the dark feeling of isolation. There was anguish at not having been able to share in the fears from close at hand; the grim days when Britain stood alone before the Americans joined. But at least there was the BBC World Service, a lifeline, and by the mid-1940s most people in the *boma* would have owned a radio. Listening to the news with friends at the club built some connection with benighted Britain. Owen was in a protected job, so he had not even served in the forces fighting in Kenya but he attained the rank of sergeant in the KDF.

But Dorothy and Owen still had not succeeded in having a child. Dorothy had lost a third pregnancy, again in the sixth month, just before their South Africa trip, and she was then under treatment in 1945 with a view to starting a pregnancy during that year.

Apart from the challenges and frustrations of work, I think that expatriate life in Nairobi in the late 1940s must have continued to be enjoyable – it was highly sociable, with dinners and bridge, sundowners and picnics, tea parties and movies. At the time there were three cinemas and a different programme every week. Dorothy noted that they saw *The Postman Always Rings Twice* ('feeble' noted Dorothy); Laurence Olivier in *Henry V* ('excellent'), *The Seventh Veil* with James Mason, Laurel and Hardy's *A-Haunting We Will Go* and Somerset Maugham's *The Constant Wife*. Admittedly, most of the movies arrived in Nairobi one or two years after their release in Europe or America, but still, it was not a bad selection. Early in 1947 Dorothy's diary is almost breathless with social life – there are many friends from the Agricultural Department (Cedric [Titus] Oates, Tom Watson and Margaret, the Camerons, Philip and Josephine Chambers, regular dinner and bridge partners, Gwen and Denzil Blunt); and also from the choir – Cicely Barton, Geoff Davies, Marjorie Hawkins, who played the organ. They met up regularly for tea, dinner, sundowners, bridge and movies.

But abruptly at the end of April 1947, everything changed in the Sunman household. Social life slowed down, perhaps one sundowner or bridge session per week, and fewer movies. At last, after months of treatment and care, Dorothy was pregnant again. The tone of the diary becomes cautious, no counting of chickens before they were hatched now. This baby was to be very carefully protected – by the end of June Dorothy was in bed most days until tea time, still on a course of progesterone injections every two days and intensive care by Dr Preston. During her enforced rest she completed two or three tapestries, some embroidery, and in the evenings Owen read aloud to her – the Dumas quartet (*The Three Musketeers* and its sequels, *The Man in the Iron Mask* – many hundreds of pages) and many of the Waverley novels – *Ivanhoe* and *The Bride of Lammermoor*). Finally, their care was rewarded – and at the end of 1947 Dorothy gave birth to a daughter, greeted with great joy. Hilary – I – was christened on 8 February at the cathedral by Bishop Crabbe.

Once out of hospital and then on to a nursing home (the Lady Northey), Dorothy and Owen leapt back into social life. On Christmas Eve they went for lunch at The Lobster Pot and on Christmas Day they had lunch at the club, and then dressed for dinner there. It must have been an exhilarating Christmas of great relief and joy – and with the luxury of a nursing home to provide baby care as and when needed.

Their house, newly built and furnished by them during 1947 – was an attractive bungalow, and they had a new dog now, Peter, a South African ridgeback with a ready temper – at least when it came to other dogs, but he was gentle with the baby. They were on the hill above Government House, their small stone bungalow surrounded by pretty flower beds and a vegetable garden that Owen had grown from seed – one of the few things Dorothy could manage during her precious pregnancy was planting seeds and pricking out bedding plants – stocks, gypsophila and salvias. They had received an allocation for a new house, off Kilimanjaro Avenue, but this was not ready until June 1948. Dorothy rejoined the choir and life was rich: full of outings with friends in Ngong and the Westwood Park Hotel and the enjoyment of dinners, sundowners and bridge.

Life with a young child in Kenya was good. There were nannies to care for the child when needed, but there were also picnics and outings, big gardens for children to play in; pictures from this period show outings with other families and groups of children playing together. There are many memoirs which highlight how good it was to be growing up in Kenya.

In 1948 they had a visit from Agnes Goble, Owen's cousin and my godmother. Travel was still long-winded and very expensive so a visitor from the UK was a rarity – but so was Agnes. Born on 1 January 1900 she was one of the many spinsters of that generation, denied the chance of a husband by the First World War. But whether she would have sought marriage indeed, I doubt. She was an independent and intellectually rigorous woman, and enjoyed a career in the insurance world – she was the first woman fund manager in the Prudential, appointed during the 1930s. Whether she would have given that up for domesticity – who knows? She was always Owen's intellectual confidante; perhaps she persuaded him that it was time to go home?

For Owen retired in 1950, after over 20 years of service. He left behind many close friends and colleagues – people he had been working with since the early 1930s. Bill Lyne-Watt who had joined the department in 1923 after the collapse of BEADOC; Stuart Gillett, his co-OWM and assistant agricultural officer in

1928 and finally his senior as director of agriculture in 1949; Cedric Oates, Tom and Margaret Hughes Rice, and Donald and Betty Storrs-Fox. The friendships one makes in the early years of working life can remain strong and lasting – and the intensity of the work and experience, the playing together, the safaris together, the sharing of humour and irritation about the quality of the housing; those first few years in Kenya must have been wonderful. Later when they were back in England they made few such close friends. The richness and challenges of the years in Kenya made a bond between those who were there which could never be achieved elsewhere.

Owen had never seen his whole life as being in Kenya; he now had a daughter whom they would want to share with family at home, and they were all too aware of impending trouble from Mau Mau. I think he felt that he and Dorothy were young enough – forty-something – to make another life back at home. Both his and Dorothy's sisters were spinsters, unlikely now to marry and have children, so the pulls of family loyalties would have been strong. And even then they would have been aware of a generation of somewhat rootless ex-colonials who were no longer at home in rapidly changing Africa, nor in the rapidly changing motherland. Owen had now more or less reached the top of the Agricultural Department tree; the alternative would be to move on to a higher posting in another colony. Many of their friends did – Tom Watson had gone to Uganda as Director of Agriculture, and Phillip Chambers – bridge player and contemporary – left Kenya at the same time as him to take up the post of Director of Agriculture in Cyprus, while Denzil Blunt moved on too. Leckie, still writing papers on livestock and wattle had moved on to a promotion in Basutoland in 1942; Lyne-Watt stayed on in Kenya and died there in 1952. Gillett became Director of Agriculture in Tanganyika and afterwards became General Manager of the Overseas Food Corporation in Dar es Salaam. Tom Hughes Rice stayed on in Kenya until the late 1950s and joined USAID for a while; the spell of Africa was too strong for him to leave and he moved to South Africa where he and Margaret ran a country club and later set up dog kennels.

But perhaps Owen and Dorothy had no appetite for this.

Dorothy took Hilary back to England in mid-1949, travelling on a BOAC Solent flying boat, from Kisumu.[42] This was a luxurious and delightful way to travel. The flying boat was the last word in luxury: definitely First Class Plus – a palatial ladies room, comfortable armchairs and Elizabeth Arden cosmetics, tables and chairs set for meals with good linen and cut glass. There were berths for reclining, breakfast in bed as the flying boat meandered gently until coming down in some suitable body of water for tea or lunch. The journey from Kisumu to Southampton took five days and entailed stops near Entebbe on the north of Lake Victoria, a night at Khartoum (passengers were put up overnight at the Empire Hotel, not the Grand!), followed by stops on the Nile at Wadi Halfa, Luxor and Rod el Farag near Cairo. Then on to Alexandria, Athens, Brindisi and Rome, before stopping again at Marseille and the final lap over France to Southampton. The journey involved three overnight stops. The Solent flying boats were relatively short lived and this journey must have been one of the last – most flying boats had ceased flying by 1940. They were designed to operate along Imperial Airways routes to South Africa and Australia and carried five crew and about 20 passengers. Luxurious they may have been, but Dorothy had an irritating 20-month-old child with her, who reportedly screamed and screamed and screamed – finally calmed by some pineapple juice. So I'm not sure that Dorothy was relaxed enough to enjoy the luxury. As to why she and I went home before Owen, I do not know.

His journey later in the year was on the SS *Madura*, the same vessel on which he had sailed to Mombasa with Dorothy, 16 years earlier. She was a 'familiar vessel to those based in Kenya – a lovely ship built in 1923 with wood panelling, a drawing room, smoking room, and the deck lounge'. The staff were all Lascars and Goans – 'Goan stewards looked after the dining room and your cabin, and the Lascars did the laundry and drew your bath for you, the baths being of boiling hot sea water and tempestuous cold water' wrote one traveller. Owen wrote regularly from the voyage, letters of hope and affection and also some uncertainty concerning the future. They had been talking of buying a farm; in one letter he is discussing what sort of car and trailer they will need to carry produce – a pig, some animal feed – to market. In another he says he is thinking of Hereford or Worcestershire.

When he left Kenya, there were dinners and farewells, formal and informal. The last night in Kenya he went out with Phil Chambers and Bill Lyne-Watt, talking with Bill on the verandah at the Norfolk Hotel until well into the early hours. On the day of departure, as ever was the case, a group came to see him off: Bishop Crabbe and Hugh Evan Hopkins; Archdeacon Leonard Beecher was there and most of the choir – including Cicely Barton who took over as Choir Leader when he left; Rosemary Gardner who had joined the choir aged 16 (her father was the chief forestry officer) and who is still alive and strong and singing in the Stroud Choir; Geoff Davies, Harry Hobkirk, Graham Hyslop and many more. His farewells from the department involved toasts and speeches; and sundowners at Government House with Cavendish Bentinck, but it was his work with the cathedral, church music and the choir that earned him the most fulsome of farewells. Leonard Beecher wrote in the *East African Standard*:

> The Festival of British Church Music which is taking place in Nairobi this weekend will mark the end of a long period of devoted service to Church Music in this country of one of its leading exponents. Mr William Owen Sunman is relinquishing his post of Assistant Director of Agriculture and going on leave pending retirement.... Wherever he has served he has made great contributions to the music of the Church. He was a chorister in Nairobi in 1933 and 1934, and thereafter was organist and choir master in turn in Kisumu, Kitale and Mombasa. In 1945, he became Master of the Choristers of All Saints' Cathedral, Nairobi, a post that he has filled with very great distinction: the services have been very greatly beautified and the whole spirit of worship enriched by his ministry of music.
>
> It was in 1947 that he enthused a small group of us to join with him in founding the Kenya Church Music Society. He felt that throughout the whole country, Christians of all

> denominations could profit from the fellowship that such a society would afford and, from a technical point of view, be greatly assisted by a musical library which the society maintains of standard works, periodical literature, music and gramophone records. From time to time, Mr Sunman has arranged festivals and conversatziones for the KCMS which have attained a very high standard, and in which school choirs as well as choirs of several churches in Nairobi, have participated.
>
> Mr Sunman's departure for Britain will leave a gap, not only for his Department (and there are others who can speak about that) but very definitely in the musical life of this country. The Church here will miss him greatly, and will find him hard, very hard, to replace. But we must console ourselves with the thought that he has led us in the establishing and maintaining standards in church music which we must endeavour to continue. We wish him, and his wife, and their little daughter Hilary, everything good for their new life in England, and we know that there is a village church, somewhere in rural England that will shortly find itself with the services of a talented Church musician.[43]

And so, good bye to Kenya for Dorothy and Owen, and on to a new and challenging life. Owen arrived at Folkestone on 23 November 1949, on an overcast rainy day, with moderate south-easterly breeze and a maximum temperature of 10°C. Not unlike the weather in which he had left for Kenya, 22 years earlier.

Part II

Echoes of Empire

7

LEGACIES

Music, agriculture and education

'There is no easy story in legacy. What is remembered and what is forgotten? There can be a chain of forgetting, the rubbing away of previous ownership as much as in the slow accretion of stories.'

Edmund de Waal, *The Hare with Amber Eyes*, p 17

'Since singing is so good a thing I wish all men would learn to sing.'

Byrd, 1588

'We must have done something right.'

Kissinger to Nigerian Commissioner for External Affairs, 1976

When Owen arrived in Mombasa in 1928, the Empire was secure, the colonial governments in for the long haul of developing and building administrations in the colonies that would, in due course, fly off as independent countries, members of a Commonwealth of Nations trained in British standards of governance and administration. When he retired in 1950 he and his contemporaries might have had much the same sense of continuity, rather reinvigorated after the war. The Colonial Development and Welfare Act of 1945 had transformed the quantity of resources available at Colony level for spending on public goods of one sort and another, a positive sign for the generation of colonial officers who had felt so poorly resourced that they were unable to achieve as much – at least for African agriculture – as they had hoped.

Yet below the surface there were significant changes. The glow of empire was beginning to dim in the public consciousness; the 1951 Festival of Britain, a festival of architecture and hope gave the Empire and the Commonwealth little prominence, unlike the Wembley Empire Exhibition of 1924. Anti-colonialism was on the rise, and nationalism within the colonies was growing stronger. The war changed expectations and attitudes in a number of ways. Nationalist movements across Asia had won a larger political role for local people, and African soldiers who had served in the British forces had seen the world and would expect more when they returned home; and these changes were more convincing as the British defeats by the Japanese demonstrated that Britain was no longer invincible.[1]

Paradoxically and perhaps in response to these external factors the post-war period under the Labour government brought a 'moral regeneration of British purpose in the colonial world'.[2] Indeed, officials after 1945 saw themselves as engaged on 'a new policy for Africa' and 'a worldwide experiment in nation building'. Arthur Creech Jones, Secretary of State for the Colonies and an ardent Fabian stated that 'the key to success lies in the development of an efficient and democratic system of local government'. Andrew Cohen, head of the Africa Division of the Colonial Office, also a Fabian and an influential thinker in the office wrote:

> The fundamental objectives in Africa are to foster the emergence of large scale societies, integrated for self-government by effective and democratic political and economic institutions both national and local, inspired by a common faith in progress and Western values and equipped with efficient techniques of production and betterment.[3]

At the same time, the Colonial Office recognised that more social services and educational facilities would have to be provided and that they had to respond to 'a rapidly increasing political consciousness among Africans'. Politics had attained a more prominent position than earlier. Political advancement of Africans – gradual, smooth and efficiently controlled – was what Creech Jones envisaged; the goal was self-government, but this was not to be hurried. A development programme was commissioned by the Secretary of State for the Colonies in 1952–53, and 1955 saw the publication of the *East Africa Royal Commission Report* which set out to provide a detailed review of the history of the three East African colonies and to identify priorities for the future.

Despite the fact that India had attained Independence in 1947, albeit with a bloody and devastating civil war, and the politics of the eastern Mediterranean was changing, independence in Africa was still seen as a long way off – many thought generations although in fact it would turn out to be little more than a decade. It was only ten years hence that Harold Macmillan made his luminous speech in South Africa, referring to:

> the strength of... African national consciousness. In different places it takes different forms, but it is happening everywhere. The *wind of change* is blowing through this continent and whether we like it or not, this growth of national consciousness is a political fact. We must all accept it as a fact, and our national policies must take account of it.[4]

Still, in the meantime there was a lot of work to be done in these nation states-to-be and the Colonial Office was prepared to take the initiative, managing change and development, rather than leaving it to governors and other officials in the field. Whitehall felt that governors had rarely taken a reforming lead – the alliance of Grigg with the settlers demonstrated this, and while Mitchell was the first governor in Kenya who saw the need to give the African population a greater role in government (at a local level), his was far from an unambiguous support for African political participation, and he had argued strongly against increasing the proportion of African representation on Legco, for example, despite his formidable record in defending native rights in Fiji where he was governor before Kenya.[5] Furthermore, the Colonial

Office felt it was better equipped for the job than others, often somewhat impatient of public opinion and pressure groups such as missionaries, but not because they did not care about human rights. On the contrary, officials regarded themselves 'as the true and efficient guardians of "the moral tradition" to which outside bodies, with imperfect access to full information could add little'.[6] This is an attitude which prevails to this day in the relations between the Department for International Development, the grandchild perhaps of the Colonial Office, and other Whitehall departments, particularly the Foreign Office and the Ministry of Defence. The paternalism in the language of the Colonial Office lingers on to the present day.

To accommodate the new priorities, the UK Colonial Service restructured itself to becoming in 1954 Her Majesty's Overseas Civil Service (HMOCS). A significant increase in the number of established posts was envisaged to support these nascent states in their growth to self-government and ultimately independence, and in the years after the war there was a major recruitment drive with a focus on technical rather than administrative staff. Paradoxically, the funding through the Colonial Development and Welfare Acts meant that in the last decades before Independence, more Colonial Office staff were needed on the ground, not fewer.[7] In Kenya, many of these recruits went out in their twenties and are alive today, with strong memories of that last decade in Kenya before Independence.

Several publications carried promotional material about Kenya in that period, seeking recruits to the Colonial Service and business activities, as well as farming settlers – although the latter were invited with greater caution than before. In 1944–45 the *East Africa Annual* was still suggesting that East Africa was a good destination for ex-soldiers – but with cautionary notes:

> One of the effects of army life is a distaste for an office life and a liking for the open air – coupled with a desire to shed discipline – but these are not enough to make a farmer, there is no harder task master than the land.... There are many people who think they can become owner farmers with a few hundred pounds, but it can cost anything from

> £2,000 to £4,000 for an undeveloped farm. Initial capital is between £1,500 and £3,000. A moderate size house with four main rooms would cost between £1,500–£2,000 to build.[8]

But what is interesting about this promotional material from the early 1950s is the positive description about Africans and African labour, which would not have been seen 20 years earlier.

> The productiveness of the African reserves would not have reached the importance that it has done without the industry and energy of the African himself. Far from satisfied with his achievements to date, after only half a century of progress, he is prepared to spend vast sums on education for himself and his children and he has demonstrated his ability to perform all types of technical work within his reach with skill and resources. You will find him occupied all over the colony as a carpenter, brick maker, builder, clerk, engine driver, mechanic, draughtsman, telegraphist, welder, shoemaker and so on. In trade, though he only recently entered the lists, he is making great strides and thousands of African shops in hundreds of markets are scattered all over the countryside. ('Kenya, Britain's most attractive colony' published by the *East African Standard*.)[9]

The tone is still paternalistic and patronising, but respect is there and recognition of the realities of development. Contrast this with what Mitchell wrote of Africans in the introduction to *The Agrarian Problem* in 1947; in this extract he too is full of respect but he is astounded at the extent to which the peoples of East and Central Africa were cut off from the mainstream of Europe, Asia and the Middle East – with whom they had the most contact through long standing trade and exchange.

> For reasons which remain a mystery today… the peoples of East and Central Africa were found in the 1890s to be in an extraordinary condition of backwardness and ignorance.

> Inland of the narrow coastal strip, they had no units of government of any size or stability; indeed, with a few exceptions such as the Buganda, nothing beyond local chiefs and patriarchs. They had no wheeled transport and (apart from the camels and donkeys of the pastoral nomads) no animal transport either; they had no roads nor towns; no tools except small hand hoes, axes, wooden digging sticks and the like; no manufactures, and no industrial products except the simplest domestic handiwork; no commerce as we understand it and no currency, although in some places barter of produce was facilitated by the use of small shells. They went stark naked or clad in the bark of trees or the skins of animals, and they had no means of writing even by hieroglyphics, nor of numbering except by their fingers or making notches on a stick or knots in a piece of grass. They had no weights and measures of general use. Perhaps most astonishing of all to modern European minds they had no calendar nor notation of time, and reckoned by the moons and seasons and the rising and setting of the sun.[10]

In fact, by the 1940s, between 20 and 30 years after the Department of Agriculture had begun working explicitly with African agriculture, the changes and level of development were remarkable. Cotton was first introduced in Nyanza Province and to a lesser extent in Coast Province, when the annual exports of cotton barely reached 800 bales. Fifteen years later, exports of cotton lint regularly reached between 55–65,000 bales per year with a value of over £150,000. There were a dozen ginneries, in Nyanza and Coast Provinces, and the cotton was of good quality. In the case of maize, there had been no exports in 1930 but by 1938–39 sales reached some 300,000 bags from Nyanza Province and more than that from Central. By 1948 this had increased by 50 per cent again, and the value of marketed maize was over £400,000. Sales of wattle bark, a lucrative and useful cash crop introduced in the early 1930s, trebled over the period. In fact, when first introduced into Fort Hall district it was much against the wishes of the people – at first the little trees were torn up. The politicians said 'oh well, this is just another plot of the white man to take your land – they just want you to plant wattle and then they will claim the land'[11] but by the mid-1940s it was understood as a valuable cash crop, and people were making a good living out of the export of bark. Yet the 1930s had seen terrible hunger and increasing soil erosion; in part due to changing practices as well as population growth.

So, 30 years after the Native Agriculture Department had been set up the outcomes – good or bad – are mixed. Was the history of the Agricultural Department – of the work of Owen and his contemporaries – one of destruction, possibly disaster, despite the best of intentions? There is never a counterfactual to actual history. The changes overall in both the African and the European agriculture had been phenomenal, but were the benefits outweighed by failure to understand the nature of the land and the nature of society? In the absence of war, would land erosion and soil exhaustion have been dealt with sooner?

Some later historians suggest that the transformation into a cash economy required by the imposition of the Hut Tax undermined traditional ways of managing life and preparing for vagaries of the weather, famine, locusts and other challenges. The sharp changes in social structure and farming practices and changes to the way land was managed, as the need to grow cash crops became more pressing, undoubtedly contributed to the loss of soil quality and erosion. According to Dr Matheka from Egerton University, land alienation caused overcrowding and overstocking which engendered over-cultivation, over-grazing and soil erosion. In the early years this socio-ecological transformation was not accompanied by technological advancement and therefore the society's margin of security against climatic variability was progressively reduced.[12] Even though the whole economy was battered by the combination of locust invasions in the late 1920s and 1930s, and the Great Depression, Matheka suggests that the local economy was less well equipped to face these external challenges than it would have been prior to the monetisation and transformation of the economy.

By the 1940s, the responses of officials – large scale removal of people, for example – spoke of desperation. There was clearly no return to the days of shifting agriculture and a subsistence economy, but could the terrible abuse of the land have been avoided, or been understood earlier? It is true that the department was spread thinly throughout the country, with most districts having only one agricultural officer supported by a team of agricultural instructors. The AOs themselves certainly felt that they were under-resourced, and under-supported by government, whose priorities had always been the development of profitable European agriculture. A different set of government priorities – focusing perhaps more on the trusteeship role of the colonial administration than on the economic (and political) strength of the European population might well have avoided some of the painful outcomes which were so evident in the mid-1940s. Yet during the 1950s rural Kenya was to some extent transformed and in 1960 Huxley could write a book entitled *A New Earth* which reports on many successes in African agriculture, even in Machakos: 'I drove for several days about the Machakos district and saw with prolonged astonishment that the... formula – terraces, grass-planting, closure to stock – really does work everywhere'. Reforms and transformation were possible; but could they be sustained?[13]

Later writers have suggested that little happened until the Swynnerton report was prepared in 1952. Roger Swynnerton came to Kenya from Tanganyika as Director of Agriculture in 1951, taking over from Gillett, and led the development of a coherent long-term agricultural development plan: The Intensification of Small Holder Agriculture in the African Areas. He had a fine reputation, and was seen as imaginative, knowledgeable and executively capable.[14] This was to lead to far reaching development and, over the period 1954–59, £8.5 million would be invested in agriculture. It has frequently been suggested that this itself was in response to Mau Mau and the Emergency – that Mau Mau forced the government to 'start helping previously neglected Africans'. As this account has shown, this view is by no means correct. In truth the Swynnerton Plan covered many projects already started but which could be stepped up given more staff and money. It also brought in ideas which had been around for some time but which had been held up by African political opposition: the core activities – land improvement, extension work, land consolidation, smallholder training – had all been in the policy canon for years. The major policy shift was to accelerate the rate of land consolidation.

Swynnerton himself admits that one of the effects of Mau Mau was to speed up the implementation of existing plans and proposals. He notes that there had been a pilot scheme for land reform in Nyeri and Kiambu districts while Tony Swan was DO, but this could only be started quietly and slowly in one village; the effect of the Emergency was to accelerate the land reform programme.[15] The agricultural teams had very good knowledge of the improvements they wanted to foster in the African agricultural sector, one of them being land consolidation – reform of land tenure – but it had been impossible to move forward on this, he suggests, for two reasons: one was the lack of finance for staff, the other was the administration's sensitiveness about interfering in any land matters.

Now in the 1950s, a new generation of enthusiastic young men began to arrive in Kenya. They came for the same reasons as had Owen and his friends back in the 1920s and 1930s: service, the desire to build better societies, to help the indigenous peoples and prepare them for future independence (for it was unambiguous that this was the aim – to build countries with strong governance structures which would take them forward to independence and a prosperous future). Adventure and a wish for outdoor life attracted many; for some the colonial background – parents, cousins – was enough to draw individuals out to the colonies; and for others there was a missionary background.

Empire directly touched many families, as sons and parents worked in the Colonial Service, but it was the Second World War which underlined the great network of nations. Troops from South Africa and Nigeria, Kenya and the Gold Coast fought not only in North Africa but also in Asia and the Middle East. One Kenyan official writes of how when he was serving in Malaya he came across an African who was wounded but could not communicate; he was Masai and the official could talk to him in his own language.

And as this new generation came to serve they brought other talents and could continue some of the wider legacies of Owen and his generation – music in particular. Richard Moss joined the Survey of Kenya as a land surveyor in 1959, at the heart of the efforts for land consolidation. He was one of a number who came and stayed a lifetime, and lived for music besides their careers (he has also produced the only current A to Z for Nairobi). John Dixon, a mathematician, came to Kenya at the same time to work on the land surveying programme. He too was a talented musician and taught himself to build and mend organs. He began his organ playing on the Mighty Wurlitzer and now, with his friend Ben Ogola, organist and choir master at the cathedral, maintains the cathedral organ. They are both still living in Nairobi, and when I met them in Nairobi in 2010 they were in the process of installing a new organ from Britain in a vast new community hall in the cathedral. Another, Phillip Dennison, joined the HMOCS as an agricultural adviser in 1956; his voyage to Kenya in 1956 was diverted by the Suez Canal crisis – one of Britain's last military stands of empire, and he found himself sailing round the whole of Africa in a voyage taking seven weeks, compared to the four or five weeks via the Suez Canal. No hardship he said; the ship was full of colleagues and girls – and brandy was 9d a shot, beer 4d! He went to work at a research institute in Eldoret, where besides work as a researcher he learned to play golf on the nine-hole golf course, and took part in music and amateur dramatics – Gilbert and Sullivan was still popular (*Trial by Jury* and *Pirates of Penzance*).

Peter Mence arrived in Kenya in 1957 and went to work in Kericho on tea estates. On board the SS *Uganda*, on his voyage to Kenya, he met David le Breton, whose mother, Peter le Breton, was a well-known musician and farmer in Kitale in the 1940s (she and Owen had both played the organ at St Luke's church in Kitale and seem to have shared the functions of choir director). Tony Davies, a talented pianist, came out to Kenya in 1963 and taught at the Prince of Wales school; he later became an accountant and joined his wife Alison's family accountancy firm – her mother was Jean Laird Bruce, founder chair of the Nairobi Musical Society (NMS). Their professions reflected the new priorities; and all these people played a role in keeping Western classical music alive in Kenya. And when they arrived none would have expected Independence within a decade.

In 1960 Dick Moss went to work in Kericho, to survey tea estates whose titles were being converted from 99 to 999 year leases. Here he met Julia Gaitskell who was born and brought up in Londiani, where her parents had a farm and they were married in 1965 in Londiani Church. Julia had graduated from the Royal Academy of Music in London and had started teaching music. Dick and Julia joined the Kericho Music Ensemble (KME), newly set up and one of the leading musical ensembles in Kenya in the fifties and sixties, a very high standard group of about 30 players. Richard and Julia between them would constitute a small orchestra – Richard plays the violin, viola, double bass, horn, trumpet, trombone, bassoon, and side drum, while Julia plays the flute, oboe, clarinet, bassoon, saxophone and piano. Peter le Breton was also a founder member. Peter Mence who had a fine counter tenor voice and Hanmer Webb Peploe (newly arrived in Kenya with Shell) joined, and over the years they played many concerts in Kericho and Nairobi as well. Richard and Julia went on to be leading lights in the Nairobi Orchestra, and Dick has been writing its history.

The KME was initially centred on the Holy Trinity Church at the centre of Kericho; and sang choral music and assembled a small orchestra. They acquired an organ from Eldoret's Dutch Reformed Church – a Walcker from South Africa with a brass plate dating from 1953 – which was found 'languishing unloved'[16] and which they repaired and installed themselves (Brooke Bond transported it in the back of one of their lorries to Holy Trinity Church). The KME played high quality music throughout Kenya, and were broadcast by the Kenya Broadcasting Corporation. One of the works was a performance of *Dido and Aeneas*, broadcast at Easter 1964; the concert was being sponsored by Brooke Bond tea and the deal was that before the performance there would be a promotional tag – 'Good music, like good tea, comes from Kericho'. Unfortunately, in error, the commercial line was not broadcast and Brooke Bond had to be reimbursed. Egerton Castle, with its mighty organ, was used for concerts and Peter Mence said that the paper rolls that Lord Egerton had used went terribly off-key and the sound was apparently dreadful. But as an organ it worked and the KME mounted a number of concerts there – at the first, in 1967, they played Handel's 2nd Organ concerto among other pieces. Now the organ is very broken and damaged.

But sadly while the tea estates flourish and Holy Trinity Church still stands, a true transplant from England complete with wicket gate, its organ has also languished, many of its keys broken; it has been replaced by a Yamaha with very large speakers. At one time the choir was affiliated to the Royal Society of Church Music (RSCM), the certificate of membership and the RSCM coats of arms still hang, faded, in the vestry at Eldoret. The church has now been joined by a striking modern one, much larger as is the way in Kenya, where new churches are built to accommodate congregations of 2,000 or 3,000 worshippers.

During the 1960s, increasing numbers of African singers joined the choirs; names remembered by Peter Mence are Elisha Sika (he has died recently), Christopher Ogoye, Joyce Ouwaya, Rebecca Tanginis, Rhoda Kamange, all of whom sang with the Kericho choir. The KME performed widely in schools throughout Kenya and made a point of performing at local schools. Moss records that in 1964 alone they performed in schools and churches in Molo, Narok, Songhor, Lumbwa, Nandi Hills, Kakamega, Kitale, Egerton Castle (this when they found the organ rolls were wildly out of tune), Nyeri and Eldoret. The KME made a point of performing at schools[17] and these regular performances drew young enthusiasts into music making and the KME put on a series of 'Concerts by Young Musicians'. (Towards the end of the 1960s the KME choir merged with the Nairobi Music Society and unnerved the existing Nairobi singers as the KME insisted on auditions. The late Roy Griffin, who told me this, passed the audition.)

Rosemary Gardner and her brothers Charles and John are of a similar generation but came via a different route. They are children of empire – in this case second generation administrators. Their father Harold Gardner arrived in Kenya in 1914 to join the Forestry Service – but soon was off to fight the Germans in Tanganyika with the Kings African Rifles in their great trail through East Africa chasing von Lettow; fighting first with horses, then mules and finally on foot. Harold was invalided out after bouts of malaria and returned to forestry. In 1920 he took home leave and got married and Rosemary was born in 1923 in Nairobi at the Maia Carberry Nursing Home. Harold became chief conservator, and they lived in a beautiful cedar-roofed house in the Arboretum that he built, near to Government House. After many years of neglect, the Arboretum has been restored and is now open to the public. Rosemary's childhood was typical for a colonial child; sent to boarding school in England, near Gloucester in her case, she and her brothers stayed there during the school holidays. There were few planes and ships' passage was too long to allow for frequent return journeys. But she had many relatives and enjoyed life – there was a real community of children of empire in boarding schools in England. She returned to Kenya at the age of 16, and both of her brothers became colonial administrators – Charles to Uganda and John a DC at Tambach.

Rosemary is a fine singer and actor, and the only person I have met who remembers Owen as choir master at Nairobi. On her return to Nairobi from boarding school at the age of 16 she joined the choir, and became timpanist for the Nairobi Orchestra. She said 'I was told that all I needed was the ability to count and the courage of my convictions' – she managed this for 20 years, and 'seldom missed a thump'. She sang with the NMS and acted with the Donovan Maule Theatre Group, set up by former professional Donovan Maule, a high quality company, which regularly did four-week repertory programmes and attracted many professionals to come out from time to time from London, often on a one-year contract. She returned to the UK in 2001, at the age of 88, but was said to have wanted to stay on for a further six months as there was a role coming up she particularly wanted to play. During the Emergency, she and the Music Society used to entertain the troops. She was awarded an MBE in 2001 for her services to music in Nairobi.

Some of Owen's contemporaries and friends stayed on for a while in Kenya. When he retired and left Nairobi and his beloved cathedral choir, the leadership was taken over by Cicely Barton until 1955 when Ray, her husband, retired. Originally a trainee manager on a coffee farm, Ray fell out with his manager and, moving to Nairobi, took up teaching, first at Prince of Wales school, a boy's grammar school at Kabete, and later at the Anglican Parklands Primary School where he became headmaster and set up a 'children's church'. He was an Anglican lay reader and later became a priest after leaving Kenya in 1955. White girls went to the Kenya Girls High School in Nairobi, and many came to sing in the cathedral choir. The first non-white girl in the choir, Cecile, had joined in about 1955, from Alliance High School.

When Cicely Barton left for the UK, the baton passed to Graham Hyslop and then to Rosemary who ran the choir for many years when she could not find anyone else to do it, for she herself preferred to sing.

The cathedral choir is still strong – stronger than before and much changed. In the cathedral vestry there are formal photographs of the choir in their robes for each year until the present. In Owen's day all the choristers were white, but over the years the pattern has changed and Roy Griffin was the last white face in the choir. Roy himself came to Kenya as an administrator in 1954 – reflecting the difficulty the Colonial Office had in recruiting in the 1950s, he noted that 'they dropped their standards and recruited people without a University degree' – and he came to Kenya instead of the University of London. He lived in Kenya from then on.[18] There are now about four different choirs at the cathedral. The NMS, now chaired by Valerie Davies,[19] is also a strong bastion of high quality choral music; they present three choral works each year, demanding music like the *St John Passion* (which David Wilcox[20] came out to conduct in 2001), or Carl Jenkin's *Armed Man*. I went to one of their choir practices held in the chapel at Kenton College – itself a strong musical centre – and was impressed by the quality of this multiracial choir.

The principal choir master now at the cathedral is Ben Ogola and the principal organist is Atigula Luvai, currently director of the Kenya Conservatoire. John Dixon formally retired as organist at the cathedral in 2011 but still plays from time to time. For some years Francis Olude, full time music teacher at Kenton College, led the choir and still shares the workload. He is a son of Marjorie Olude Macgoye, who had married a Luo doctor in Kenya in 1960 – a rare occurrence, although not unheard of – and became a well-known and much respected writer. The only child of a working-class family in Southampton, England, she moved to Nairobi in 1954 as a bookseller for the CMS. She was, according to Kurtz, colleague and co-author, apparently always something of a maverick. She left the CMS when she married and turned to writing; she became close to his Luo extended family and her rich and beautiful writings reflect the complex relationships between Luo and European, and the changing nature of Kenyan society. Following Independence in 1963, she took Kenyan citizenship, and raised four children in Kenya, all musical, one of whom is Francis.

The post-war vibrancy of classical music in Kenya rolled on into the early fifties, and the quality of the Kenya Conservatoire which had been established in 1944 by Isherwood, Fryling and Kofsky attracted many well-known – even great – artists. Names included Vladimir Ashkenazy, John Lill, John Ogden, Vlach and the Chilingirian Quartet and the Berlin Philharmonic. At the time, Nat Kofsky was the director and by 1960 the conservatoire had a staff of 24 teachers and an average of 375 pupils of all races.[21] He initiated a building fund and the conservatoire was able to construct a purpose built space in the new Kenya Cultural Centre, opposite the Norfolk Hotel. The building consisted of a corridor with 12 teaching and practice rooms, a suite of offices and a ballet studio – and indeed it still does (2010). Membership rose during the 1980s to over 600, when the conservatoire became responsible for administering the Associated Board of the Royal Schools of Music practical examinations. Nat Kofsky remained director for many years, and the conservatoire's fortunes fluctuated. Financially they were always stretched; they sought Kenyan musicians and kept the fees appropriately low to ensure an inclusive programme, but even so, of the 375 pupils only about 45 were African. Many of the African places have been funded through annual bursaries from the British Council.

Luvai says that from the 1950s through to the 1970s there was always quite strong classical music in Nairobi, but it was mainly among the expatriate, European community, 'pretty exclusive obviously',[22] and it has taken time for Western style classical music to find a place in wider Kenyan society. In the 1970s, music was taught in some Kenyan secondary schools by people like Julia Moss, who has long been a teacher at the Kenya High School, teaching Kenyans – 'the unglamorous end where all the work is done' as she puts it, but Luvai feels that the students only achieve the lower grades – 'if you start the piano when you are 16 you don't go very far' he says. Interestingly, when there was an influx of Ugandans to Kenya in the 1970s and 1980s (post Idi Amin) they seemed to be much more accustomed to Western classical music and for a while kept music and theatre alive. He thinks that there was more cultural transfer in Uganda than in Kenya. There was an artistic flowering under the influence of the Ugandans at this time, but as they moved on or returned to Uganda as the political climate improved, it withered, although it left some influence on Kenyan children who had seen their Ugandan classmates taking music seriously and playing instruments.

The first African Director to the Conservatoire, Christine Mutungi, was appointed in 1994; Luvai was appointed in 2002 and according to Dick Moss 'has been steering a wise and fruitful course'. He has founded a conservatoire orchestra – equipped with second hand instruments through the offices of Valerie Kent (now Davies) who scoured the UK for second hand instruments, which the RAF flew out for them free of charge. Luvai has instituted termly pupil concerts too, which attract both parents and friends of the players – they play a mix of classical pieces and more popular arrangements, to keep the music accessible to new audiences. By mid-2009, the orchestra had completed tours to Kampala and Dar es Salaam, with more planned. As Luvai says: 'Students acquire the taste and after a while classical takes over from the pop, so they join the orchestra, take on bigger challenges, and go on to the Nairobi Orchestra.' 'In its sixty-fifth year, then, the Conservatoire was in good order' says Dick Moss.

There is also an appetite for classical music – particularly the piano – from the Asian population perhaps reflecting the strength of western music in Asia, particularly China and Japan but also India. Atigula comments that for the first time there is hope for posterity, now that locals are in it, because they are not going anywhere.

In 1964, Graham Hyslop, by then organising inspector of music and drama was asked to prepare a report on 'The Prospects of Music in Education in Kenya'. Although there was little formal musical education in place in schools at that time, Hyslop points out that at the time there were various departments and organisations which were contributing music at school level – such as the Nyanza Musical Society which, during the late 1940s, ran a course for choir masters with the aim of raising the standards of music in schools – about 80 teachers attended each year. At the time he was writing, the music education was totally focused on singing, but often of a very simple kind and certainly at the expense of instrumental music. He writes passionately of the inherent value of music in reflecting the soul as well as the intellect of humanity. Where there is music – he mentions Alliance High School, Kenya High School and various other schools – musical activity has centred around the 'School Choir and its participation in the Annual Music Festivals. This is fine, but it is not enough'. The rich musical heritage of Kenya deserves more than this, he writes, and 'this inheritance unquestionably demands its rightful place in the educational programmes to be worked out in Africa'. He feels passionately that any debate on whether musical education should focus on Western or African music is sterile.

> One group maintains that Western music is so far ahead of African music in melody and harmony that it should be the music to be taught in schools. Another group holds the view that in a sovereign African state, nothing but African music should be taught.... I cannot support these extremist views; Western music and African music should be taught side by side, because each has a distinct contribution to make to the life of a school and the musical education of the pupil.[23]

Although music teaching in schools never materialised as he would have hoped, the choral tradition in Kenya today owes much to Hyslop – the use of a traditional melody with a pentatonic harmony, in the traditional musical structure. Many of his pupils – particularly, Arthur Kemole, Washington Omundi, and Philip Auma – are now teachers and themselves write many arrangements of choral music.

One Sunday I went to the morning service in the Mombasa Memorial Cathedral, a long choral communion service and a packed congregation. The service was a mix of a contemporary modern musical style and a more traditional formula, which was beginning as I arrived. I heard the notices, the banns, and then the sermon, followed by communion. It was easy after a while to imagine the white choir and the mixed congregation of 60 years ago, the same hymns and liturgy, the same voices and the same prayers. Although totally changed in the sense that the congregation was completely African, and the choir vestments were vivid Trinity green, the focus of the sermon (the importance of work and the need to avoid idleness) could have been preached by any Protestant pastor over the past 150 years. And indeed the words of the prayers are hardly changed from the great 1662 prayer book – modernised perhaps but enough of the old cadences ring through the years. This continuity makes some sense of the appeal of the Church to so many; the extraordinary feeling that the individual is part of a worldwide movement with shared goals, values and history.

The cathedral now has an old electric organ, and there is a group of elderly Kenyan musicians – Philip Angore, the organist and Joseph Kazungu, the present director of the choir, who are fighting to keep the traditions of classical church music, and the playing of the organ. After the service at the cathedral, I talked with Angore, and Joseph Kazungu. Now in their 60s or 70s, both are Hyslop protégés. Joseph is 61 and very feisty in defence of formal music; and they in turn have a protégé, Zachary Njohn Kinyua, a youngster who plays the electronic organ but is learning the real wind organ. It is very difficult he says; using the feet as well as the hands, and two keyboards too. When asked how the music has changed these veterans spoke of the local chanting with beats for dancing, and how the standard music repertoire is disappearing. Music books have been banned in some churches because no one knows how to use them or to read music. Is it a loss? 'Yes, yes, we feel very strongly about it.' The tragedy is that there are now few competent organists in Kenya – most are former Hyslop pupils – so that Zachary is seen as a promise for the future.[24] But Luvai suggests that the problem with many choirs in Kenya is that while they sing the Hyslop type of music very well, they hardly sing the classical repertoire at all. The only choir whose members read music is the cathedral choir, where they do a full sung evensong once a week to meet the needs of the Anglican liturgy.

The legacy of choral singing has shown far greater fortitude over the years than instrumental playing. Yet perhaps this is changing. There has been a core of people keeping Western music alive in Kenya; first the KME and then the Nairobi Orchestra have maintained a high standard of music making over the years and now growing interest among the African and Asian communities. There is a regular classical music show on one of the commercial radio stations, hosted by musician and impresario Elizabeth Njorogo.

But if the musical legacy is somewhat attenuated, the agricultural legacy is strong. The need for a transformation in African agriculture had long been recognised by Owen and his colleagues, and, finally, some of the policies formulated by the Department of Agriculture after the war were put into place in the early

1950s, with the injection of funds which followed the publication of a key report by Roger Swynnerton, the Director of Agriculture who succeeded Stuart Gillett. The central tenet of the Swynnerton report was:

> Sound agricultural development is dependent upon a system of land tenure which will make available to the African farmer a unit of land and a system of farming whose production will support a family at a level, taking into account perquisites acquired from the farm, comparable with other occupations. He must be provided with such security of tenure as will encourage him to invest his labour profits into the development of his farm and as will enable him to offer it as security against such financial credits as he may wish to secure from such sources as may lie open to him.[25]

This – the consolidation of many scattered smallholdings into larger economic units with clear title – was the heart of the policies of the 1950s, and had really got under way in response to the Emergency.

Dick Moss' description of his work gives a clear idea of the work involved in building the new Kenya. The department produced topographic maps for general development, detailed plans for title registration and all the various surveys in support of the land reform processes getting under way in the African Reserves at that time. At one point, Moss was posted to Nyeri as provincial surveyor Central Province which, as he says:

> was the hot-house of Land Consolidation. The actual adjudication of ownership of the land fragments was undertaken by staff of the Land Consolidation Dept. and we just plotted the spatial information in the right place, more or less. My parish included Fort Hall, Nyeri, Kirinyaga, Embu and Meru districts and progress was made in that direction.[26]

The consolidation process involved a survey of the land, identifying land ownership, assessment of the acreage, measurement of individual fragments, allocation of a certain proportion of each individual's land to common goods – hospital, roads, schools, etc. Individuals had a 30 day period in which to make complaints after which each owner's land was assessed and demarcated, and the owner received a registration certificate.

The aim of the process was to achieve a subsistence living plus cash earnings of £100 per year, and the 'economic' size of a holding was estimated at 7.5 acres.[27] Most importantly, for the first time African farmers had access to credit. Consolidation began in Central Province, and was able to build on the policy of 'villageisation' of the Kikuyu people during the Emergency; and consolidation was presented to the Kikuyu as a reward for loyalty to the government. The Kikuyu were ready to accept consolidation for most of them already desired secure title and they quickly recognised the advantages of consolidated holdings.

The main programme had been completed in Central Province by 1959, and by 1963, 1,272,000 acres of land had been registered at a total cost of £3.2 million. 'Consolidation became – not just a programme – but a creed'[28] and in 1962 an opinion poll indicated that the majority of Africans were themselves in favour of reform.[29] The value of output from the small farm sector increased dramatically between 1955 and 1966: in 1955 the value of marketed crops was £5.1 million while the value of subsistence crops was £48 million, but by 1966 the value of marketed crops had increased four-fold to £19 million, and subsistence crops too had increased by nearly 60 per cent to an estimated value of £75 million. The Swynnerton report 'sought to raise the productivity of the African lands, its human and stock carrying capacity, the income and standard of living of the people, while at the same time effecting a substantial increase in the resources and economy of the Colony'. But by 1965, of the 114,300 consolidated farms in Kikuyu, only 15,800 were 7.5 acres or more, so while there were benefits they did not, in the main, arise from more economic plots, but perhaps from security and access to better inputs – and credit.

Between 1978 and 1984 Oxford University carried out a review of projects undertaken in Africa in the colonial period. Economist Anne Thurston led an examination into the development of smallholder agriculture in Kenya post Swynnerton. In undertaking her research she interviewed many of the agricultural officers who had been working in Kenya in the 1940s and 1950s, most of whom had been working in Kenya in the 1940s and some earlier.[30] Thurston concludes that although some officials hoped that the Swynnerton Plan would achieve miracles, it was not actually designed to do so. But it was a realistic agricultural programme based on field experience and this, she argues, lies at the heart of the programme. Population pressure and poverty due to land shortage could not be overcome by the plan, but the levels of growth in cash cropping – coffee, tea, pyrethrum – met or exceeded plan goals and transformed lives. The plan had succeeded in shifting the emphasis in Kenyan agricultural policy from large scale European farming side by side with African subsistence agriculture to commercial peasant agriculture.[31] Later, in the late 1950s, the government removed the racial restrictions in the Highlands and permitted resettlement into new areas for Africans, thus relieving the pressure on Central Province and providing space for the tenets of the Swynnerton Plan. Thurston concludes that without the plan, production would not have increased so rapidly or at such a high standard.

It is worth quoting some of her conclusions in full. She writes:

> The agricultural history over a 30 year period illustrates aspects of the colonial development and development generally. While the broad lines of policy were discussed in London and Nairobi, the content of development in specific parts of Kenya was evolved by the men in the field. When political restrictions at the centre prevented them from being able to either extend the land unit or offer economic benefits, they were unable to work effectively. When they could offer incentives, with technical and administrative back up, they achieved high standards rapidly.... Perhaps the most crucial lesson to be drawn from the Swynnerton

> Plan is the importance of the field officer's role in development. Local solutions do not solve national problems, but when Swynnerton was able to draw them together and amplify them, with backing from the centre, there was a breakthrough. But the Plan succeeded in a way that present international projects often do not because it was rooted in the field.[32]

It was grounded in the work that the agricultural officers had been doing over 30 years.

By 1960 Huxley could write in *A New Earth* with respect and confidence of the new agriculture in Kenya, albeit with a touch of caution. She travels all of Kenya and discusses 'Progress and Change'. Progress, she says, may be a giant or an angel or an ogre, but at the time she saw he was striding through Africa at breakneck speed. 'One can neither halt Progress, nor be sure that because he is walking forward he approaches a goal. He may be going round in a circle, or back to camp. The point is, he is on the move'. And she is writing of changes that are 'revolutionizing the lives of Africa peasants'. She writes:

> Each tribal group still holds jealously to its traditional region, even if it reached its present homeland only in the last few hundred years or indeed within the last century. Kenya is honeycombed with land barriers, both of race and tribe; but whether people like it or not (and they seldom do) Progress in his headlong match is knocking them over. This year, the practice which for half a century has reserved the white highlands for farmers of European descent has ended, and in this region no racial bar now inhibits the lease or sale of land.[33]

Incidentally, she points out that a bar still prevents the exchange of land between members of one tribe and another.

She writes of the different regions of Kenya and the different patterns of development and change – often characterised by the nature of the tribe. The Kikuyu have always been the quickest while the Luo are thought of as stubborn and less responsive to

'Progress'. She too acknowledges the importance of knowledge on the ground. Working with the existing highland tea industry and committed agricultural advisers, the tea industry took root (although one successful project in Kipsigis country meant the clearing of one of the last surviving areas of native forest). Kenya is now, in 2013, the largest exporter of black tea in the world, through a mix of plantation and smallholder tea estates.[34]

The contribution of the agricultural advisers is being recognised now in Kenya. Writing in 2010, a young Tugen academic, Daniel Kandagor, puts the case that the colonial period effectively transformed society by the introduction of modern and much more productive agriculture. The transformation of the Tugen, a remote and 'primitive' tribe, into modern farmers over the first half of the twentieth century was due to a combination of factors, training and interaction with education and the missions. For example, while better maize was being introduced throughout Nyanza and other Native Reserve areas, the Tugen refused to grow it. But during the 1930s a number of Tugen went to train in agriculture at Bukura and 'after training, [the men who had been trained] implemented new systems of land use, crop rotation, allocation of labour, irrigation of crops, and ridging contours by planting kikuyu grass'. Adoption of modern agricultural methods went hand in hand with education – and the local missions were essential. Demonstration plots and later irrigation contributed to the transformation. Kandagor also suggested that after the war, Tugen war veterans 'who had been exposed to the world abroad practised what they had seen while in the war'. He also comments that in the Emergency of the 1950s:

> the Tugen benefitted a great deal, particularly in the agricultural sector, partly because Kikuyu internees were used to build canals at the Perkerra dam [which directly affected the Tugen] and second, by interacting with the hard working agricultural Kikuyu, the Tugen learned the value of hard work – the hard work and efficiency of the Kikuyu provided a demonstration to the Tugen.[35]

Indeed, now, 50 years after Independence, it can be seen that the agricultural structure of Kenya is built on the pioneering and experimentation of the settlers and the efforts of the government to develop a stronger agricultural base for the African population. Those years of experimental plots, improved seeds, exhortation and *barazas;* the efforts and commitment of Owen, Bill Lyne-Watt, Tom Hughes Rice and the others laid a foundation for the future. In 2010, agriculture contributed 20–30 per cent to the economy of Kenya and of exports, and is based on tea, coffee, cut flowers and cereals. In 1950, the pattern was recognisably the same: cereals, coffee, pyrethrum – tea came later. Agriculture in the colonial period was based first on what would grow and then whether anyone would buy it. Today there is more agricultural policy, but it turns out that the crops and products identified in the colonial period are still the backbone of Kenyan agriculture – and agriculture is still the backbone of the Kenyan economy. Sir Richard Turnbull, Chief Secretary in Kenya during the Mau Mau period, understood the significance of the Colonial Agricultural Programme:

> Nobody could fail to be impressed with the quality of the men that headed the Dept of Ag – it was generally accepted that the Ministry had at its service some of the most brilliant men in that particular field in the whole of the commonwealth. As for the activities of the department, how could one describe them better than to say that the economic future of the new Kenya depended more on the policies devised by the ministry and the efficiency with which they were put into execution than upon any other factor.[36]

But could they have done better? In the introduction to this book I suggested that this is a contested history and indeed it is. I have tried to show that at the very least the Department of Agriculture in the period between the wars pursued a policy of development and strengthening of African agriculture, albeit with very limited resources, but with a strong commitment to betterment of the African lot. The exact amount of available resources devoted to

African agriculture is not clear, but analysts writing later talk bitterly of the destructive, coercive nature of agricultural policy in the Reserves. Historical analysis needs distancing tools to understand the wider arguments, but in terms of the work Owen and his contemporaries were trying to do, I doubt that they would recognise themselves in this, a quote from a Marxist historical analysis of 2000:

> My intention in this paper is to examine the colonial narratives about agriculture and the environment in Kenya in the 25 years between 1920 and 1945. This was a period that saw *the increasing reliance of the state on African agricultural production from the Depression onwards* [my italics] and growing evidence of environmental costs of policies to expand such production.... From the early 1920s, the doctrine of 'trusteeship' which informed colonial thinking, drew increasingly on the authoritative claims of 'scientific method'... to generate discourses of 'betterment' and 'environmentalism' which legitimated a deepening of administrative control.[37]

Owen and his colleagues would have seen themselves as trying to develop African agriculture and the notion of 'trusteeship' was to strengthen and build up African agriculture – with, always, the idea in the background that the final aim was to build a country ready for independence – very different from the later interpretation.

Still, there are legitimate criticisms of agricultural policy of the period. The same writer draws attention to the role of women in African agriculture and it is doubtless the case that this was not well understood. For example, Alex Holm, when Director of Agriculture writes in an overview paper about agriculture in Kenya in 1926:

> the problem [of increasing output] will only be solved when every male native who is of working age and is also physically fit will work regularly throughout the year, either on his own account or that of his employer. An

> undue share of the work involved in native production is, under existing native habits, dependent upon the efforts of the women folk, who might be advantageously engaged on greater service in the work of the home and the care of children thus reducing the heavy mortality which at present exists among child life.[38]

The failure to understand the role of women in traditional African agriculture is undoubtedly a serious weakness in the colonial agricultural project. In former times of course the responsibility of the men was as warriors – now redundant in the wake of the effective *pax Britannica.*

Such a bias in agricultural advice has been long lasting. Many years later when I studied agricultural economics and development studies, writers were drawing attention to the fact that agricultural extension services were normally addressed to men rather than women, regardless of the realities on the ground. The combined failure to understand the realities of smallholder agriculture and the desire to introduce a 'modern' model of plantation and large scale farming, run by men to the exclusion of women, has been remarkably persistent. It does seem logical from a European perspective – not just because in Europe men would tend to be farmers and cultivators while women on the farm would be responsible for horticulture, small animals and for food preparation and preservation – but also in traditional European societies it is the men who are the breadwinners and go out to work rather than the women. Men in Kenya indeed took up estate and plantation farming steadily once introduced, but in my view the importance of women's agriculture in basic food production has been underestimated over the years. In this I think the department may well have been at fault, although it would seem totally legitimate to male agricultural advisers to focus their efforts on getting men to work more effectively. But in practice the function of the agricultural inspectors and the local national councils would have been to bridge the gap in understanding. Indeed, I had discussed this with Owen in the early 1970s, when I was studying development economics for my Master's degree. He argued that in fact the agricultural officers were very aware of local practices – but nonetheless wanted to change them to improve output and incomes. In the 'development business' that is, to this day, the name of the game – to change, improve, accelerate, strengthen, reform – but sometimes this means riding roughshod over existing practices.

Nonetheless, in my view it is clear that during the 1930s and 1940s much was done to put in place a robust and lasting agriculture; not just the growing of crops, but through education and training. I have written mainly of African agriculture and the efforts of a small but committed group of men to strengthen and improve it. But the settler agriculture also paved the way for the currently strong agricultural sector – pioneering with crops and styles of cultivation – and experimenting and investing, building knowledge among the land owners and among Africans that worked on the farms. Many of the settlers were hard working men and women endeavouring to develop farms and businesses in pioneering conditions, but their voice is sometimes hard to hear above 'the trumpeting of the self-appointed leaders'. The transfer of knowledge was not particularly planned – indeed many of the more intransigent settlers were reluctant to enhance the capacity of their African workers for fear they would challenge the status quo (although by ignoring African wishes they reinforced the suspicions and tensions that underlay rebellion in the 1940s and 1950s). But there is no doubt that their efforts – together with departmental efforts in experimentation, training, development of seeds, raising standards and introducing better techniques – made a central contribution to the present day agricultural sector. At the moment of Independence in 1963 Kenyatta made a profound and moving speech recognising their contribution:

> I am a politician but I am a farmer like you... I think the soil joins us all and therefore we have a kind of mutual understanding... If we must live together, if we must work together, we must talk together, exchange views... we must learn to forgive one another. There is no perfect society anywhere. Whether we are white, brown or black we are not angels. We are human beings and as such are bound to

> make mistakes. But there is a great gift that we can exercise, that is to forgive one another. If you have done harm to me, it is for me to forgive you. If I have done harm to you it is for you to forgive me. *I want you to stay and farm and farm well in this country.* All of us, white, brown and black, can work together to make this country great... let us join hands and work together for the benefit of Kenya and not for the benefit of one particular community. *This is what we beg you to believe...* we are going to forgive the past and look forward to the future. Because if we start thinking about the past, what time shall we have to build the future?[39]

The motto of Independence was '*harambee*' – which means pulling together. Yet P. D. Adams, the deputy director of the settlement programme, articulates the anxieties of the times: the white man was asking 'are we staying or leaving? Can we trust them?' while the black man's questions were equally 'Can we trust them? Is it really true? Will we get Uhuru?'[40]

Some white farmers did indeed stay on, but after Independence they were not permitted to retain British nationality as well as Kenyan. In the early 1960s, following the landmark Lancaster House Conference[41] the British government together with other funding agencies supported a programme to buy out white farmers and resettle the land with smallholder African farmers. At the time of Independence European farmers – about 3,000 of them – held some 7 million acres of land in the White Highlands. Under the settlement schemes, about 2 million acres was bought from the white owners on a 'willing buyer–willing seller' valuation and the land which held about 800 European farms was transformed into 35,000 smallholder farms. One of the main aims of this land transfer programme was to introduce African farmers to the advanced systems of production of the Highlands. Other objectives were political, of course – to dilute the white concentration in the Highlands – but also social, to settle on the land 'people who had previously been landless and had been unemployed, or living and working where they could in African areas, or who had worked as employees on the farms taken over for resettlement'. The total cost of the settlement programme was £25 million and it was a formidable undertaking – to 'turn a large number of inexperienced people into successful, modern, cash-oriented farmers in a short time is a very difficult educational and extension programme'.

In his report, effectively the official account of what happened, Adams writes:

> Kenya has been fortunate in having European farmers who gave and are still giving their valuable and vital contribution to the whole agricultural industry for many years before and after Independence. In spite of faults and wrong attitudes, however, the pioneering spirit and the toil of hard working European farms has laid lasting foundations of sound farming methods in Kenya.[42]

However, while the scheme aimed to set up a mechanism whereby European farmers wishing to sell up could expect to be paid a reasonable price,[43] in practice one of the longstanding grievances with the programme, from the perspective of the white farmers, was that the valuation on the farm house was capped at £1,300 – it was deemed that the ex-European farm houses were of too high a standard to be used by the incoming new African farmers, and higher values would only be paid for the houses if the planners could foresee that the building could be used in the future for community purposes.[44] Not only that, the price used for valuation of the land was based on Kenya land values of 1959, which would not enable the outgoing farmers to make a new start outside Kenya, certainly not in England. In the words of Adams:

> the price limitation on homesteads was undoubtedly a cause for much disappointment and bitterness to the outgoing European owners who in many cases had spent years of care building up their homes – often from very basic beginnings – and developing beautiful homes and gardens. Meanness does not make friends.[45]

But it is highly unlikely that the political mood in the UK would have tolerated generous payments to the white settlers.

In a book of essays, Chinua Achebe, the great Nigerian writer, reflects on the impact of colonialism. He tells of how, in 1976, Kissinger spoke to Joseph Garba, Nigerian foreign minister at the United Nations, and asked him what he thought America was doing wrong in Africa. 'Everything', replied Garba. Kissinger's next comment was to say 'Statistically that is impossible. Even if it is unintentional, we must be doing something right'. That exchange, writes Achebe, could easily have been about colonialism.[46]

Even Achebe conceded that the British had left the profoundly transforming legacy of education. Begun informally through the mission schools, many Africans quickly saw that this was one of the tools with which they could emulate the white man and his security and wealth. By the 1930s and 1940s there was a generation of Africans who saw the value of education – including education for women. When I was in Meru, I met the daughter of one of Owen's agricultural instructors, Asenath Kaimuri Nyamu. She told me that her father had so appreciated the education and training he received as an agricultural instructor that he made sure that she, Asenath, should be educated, even though her mother was not an educated woman. Asenath went to the local mission school and now, as head of the Rural Community Development Agency, an NGO, is dealing with other quite different problems in Meru: supporting women with HIV or who have been abused as a result of local preoccupation with *meera*, a leafy and quite addictive stimulant (known in the Middle East as *khat*).[47] Education is a liberating force, and particularly so for women. Wangaari Matthai, Nobel Peace Prize winner and campaigner for environmental issues in Kenya, writes in her autobiography of her delight when her father permitted her to go to school and they found the money for her as well as her brothers.

Kenda Mutongi,[48] a Kenyan academic at the University of Chicago who was born in the 1950s, questioned why, when she was growing up, her parents had referred to the colonial period in positive, even glowing, terms, whereas her teachers were furiously anti-colonialist. She explores the question by examining how widows are treated, under both colonial and post-colonial rule, and their access to justice and land. Widows are an interesting group to examine since their situation is not particularly rare – about 30 per cent of families in Africa are headed by widows – but their treatment under law and in society provides, Mutongi suggests, a good barometer of changes in village life, or changes in the economy, or social and familial circumstances. She found elderly women of the Maragolis (her own) tribe of Western Kenya and interviewed them and their relatives, and her mother, herself a widow since Mutongi's childhood. Their responses suggested that, yes, perhaps the decisions made under the DCs were fairer, less influenced by family and peer group pressures and poorly thought through legislation than was later the case. She heard a consistent message: 'We had schools and hospitals and white people took good care of us. The white teachers were serious about teaching – not like the teachers of today who take naps in office and never show up for class'. 'White people never deceived you; there was no corruption like we have now among our own people.' But in her concluding chapter, she (Mutongi) cannot quite believe this: 'perhaps the old women were too old to remember correctly'. As she says:

> what I had anticipated was for the Maragoli men and women to tell me about the evils of colonialism, as my high school teachers had done a decade earlier... and speak of the callous immorality of colonial rule, as all those scholars had verified over and over again.[49]

Yet she finds that her radical former teachers too had changed their views: 'life is definitely worse than in the old days' said one; 'our leaders have failed us', sighed another. 'Life is very hard in Kenya these days; the white people treated us better' said one lady, and another former teacher said 'Our leaders don't care about us anymore; at least white people told the truth and were not corrupt'.

Mutongi found that the rosy hindsight for the old days was strongly coloured by a feeling, a perception that the contemporary politicians had failed. She suggests, probably correctly, that some of the nostalgia would be due to the exaggerated expectations of post-colonial bliss at the time of Independence, but that this had combined with the failure of leaders to meet the needs of ordinary citizens. She concludes that 'many of my informants were at least partially motivated to praise the colonisers because they wanted to see change in the contemporary government, not necessarily because they had liked colonial rule'. But rarely did she hear anyone express confidence that there were institutions or administrators that could be turned to for assistance.

Dispassionate administration and governance was perhaps the other important contribution of the British colonial system; perhaps Kissinger was right: 'we must have done something right'.

Whether it is ever right to colonise, to expand into others' countries and lands and settle, assuming rights of ownership and dominance is indeed questionable. For my generation, it has carried a powerful shadow of shame. But in the nineteenth century, there was not a corner of the world untouched by the energy and economic power of the European industrial revolution. Searching for goods, for raw materials, for adventure and for Christian souls, the European powers were confident of their abilities and religion and – most importantly – the rightness of their cause. In the early twentieth century empire was seen as a source of peace (perhaps even then this was wishful thinking), the commonwealth of man, the sisterhood of nations, the inspiring visions of Lord Reith. There was inevitably a sense of superiority and paternalism – it could be said that the colonialists treated the natives as children when they were compliant and as terrorists when they were not. But who could help but feel superior when they have the internal combustion engine, literacy and science, and are contemplating people who do not even have the wheel? (The colonialists also had overwhelmingly more effective weapons – the Gatling gun versus the spear). The problem is that there is no counterfactual.

Sir Richard Turnbull famously said, when asked by Denis Healey 'tell me Sir Richard, what are the enduring legacies that Britain will leave to Africa?' 'Association football', replied his Excellency, 'and the expression f*** off'. But there was more than that, and Turnbull and even Achebe saw that education was the great legacy. Education was seen early on as the necessity in order for people to change and acquire the goods and riches of the immigrant caste. By the 1920s and 1930s Africans sought to build their own schools, and families saved hard to pay school fees. Politicians – Jomo Kenyatta, Odinga Oginga, and Nobel Prize winner Wangari Matthai – all were taught first in mission schools. Chinua Achebe, N'gugi wa'Thiongo of Kenya, and Dominic Mulaisho of Zambia, great writers all, learned their literacy in mission schools. Washington Omundi and Luvai Atigula – carrying the torch of the best of European culture, music and art – learned through the Christian churches and church music.

But the view in contemporary Britain, during the final years of empire was of a sense of embarrassment, and shame. Political crises such as Suez reinforced how pathetic empire had become. In part that is what this book is about: to tell the story of this forgotten generation whose intentions were good – but who have been misunderstood.

8

REFLECTIONS

Children of empire

> 'I was one of a lucky generation, which avoided the ambiguities of Empire and the horrors of war, setting out in the 1960s at a time of idealism and optimism. Thanks to the policies of Mao-Tse Tung and Pandhit Nehru the West retained its predominance throughout my working life.'
>
> Jeremy Berkoff, *Incidents in the life of a development economist*, Preface

By the 1960s the old pattern of colonial service had gone. During the 1950s Whitehall had envisaged a central pool of experts to support development but this had not materialised. In the 1950s people might have joined the HMOCS with a reasonable expectation of a career and pension, but by the early 1960s it was clear 'that recruitment [to overseas posts] would now be based on the principle that service overseas would be for limited periods and as part of a career structure based primarily in Britain'.[1] There would no longer be a regular establishment of overseas posts; rather a series of contracts or loan appointments to deal with development and welfare issues. The 1960 government White Paper separated the traditional government role in the colonies from aid and development programmes for the first time, and in 1961 the Department for Technical Cooperation was set up, with responsibility for recruitment for the overseas public services. Its function was to 'bring together the expertise on colonial development previously spread across several government departments'. Thus, the Colonial Service had moved on to a programme of technical support, while the HMOCS provided governors and administration staff until independence in the various countries.

To quote the brief history of the Department for Overseas Development on its website: 'Under the Labour government of 1964–70, the Department for Technical Cooperation was combined with the overseas aid policy functions of the Foreign, Commonwealth Relations and Colonial Office' into the Ministry of Overseas Development (ODM). In 1965, a government White Paper stated that the UK had a moral duty for development and that development was in the nation's long-term interest. The term 'foreign aid policy' had not been heard until that moment, but it is the area into which the technical support provided by the Colonial Office metamorphosed.

Thereafter there was a regular tussle between Labour and Conservative governments as to the role of the development programme and the relations with the Foreign Office. In 1970, when a Conservative government was returned, the ODM became the Overseas Development Administration (ODA) and was planted firmly under the Foreign Office, accountable to the Foreign Secretary. In 1974, returned to power again, the Labour government announced that there would once again be a separate Ministry for Overseas Development with its own Minister of State. But in November 1979 the ministry again became the ODA, a functional wing of the Foreign and Commonwealth Office, and it would remain in that position until the landslide Labour victory of 1997, when the Department for International Development (DFID) was set up as a separate government department with, for the first time since the 1960s, a minister sitting in the cabinet.

If political commitment to aid and development is measurable by the political status of its implementing agency, then the continuing Labour support for aid is a natural descendant of earlier preferences. Hyam[2] comments that the Colonial Office saw itself as humane and progressive, unable to identify with extreme right wing views of empire. They were proud of empire but sceptical and they were happiest and worked most effectively under radical administrations – the Liberal government of 1905–15 and the Labour government of 1945–51 – and indeed, much later, under the Labour government of 1997–2008.

The 1945 Colonial Development and Welfare Act can be seen as the UK's first real aid commitment. In 1946 France followed a similar pattern and established her own 'aid' body, the Fonds d'Investissement pour le Développement Économique et Social des Territoires d'Outre-mer (FIDES). The bilateral aid programme is clearly a legacy of the colonial era, and provided the context for many of the later children of empire who carried on in the tradition of working overseas with a sense of commitment and wish to do well by the recipient countries – not exactly a 'robust Christianity' in these more atheistic days but certainly many of those who have worked in this field have had a strong sense of altruism and a commitment to 'do good'. There is also, undeniably, a sense of adventure and wanderlust among the aid working fraternity which echoes that of earlier colonial days.

It is a world I have worked in since 1974, after studying Development Economics at the University of East Anglia which is occasionally thought to be a hotbed of left wing thinking. It is certainly the case that the teaching and thinking of the day had little sympathy for the colonial efforts; the radical analytical Marxist language which I have quoted from time to time in this book was the language I trained in and, while I never felt that the motives of my parents or their agricultural colleagues were anything but benign, I saw them as misplaced, that they should not have been there in the first place.

So why did I go into this field? History casts curious shadows and I suppose my motivations were a combination of curiosity about life and the romance of Africa – as I have said I grew up with stories of Kenya surrounding me. But another motive must have been to redress the balance, to try to undo the malign legacies of empire.

It can be no coincidence that among my colleagues over the years – in the 'development business' as it is now called – like me, many are children of empire with fathers in the Colonial Service. I conducted a very informal survey of a society of international economists that I belong to, to see how strong the empire inheritance is.[3] The answer is – very. One economist friend of mine who first went to Kenya before Independence is the daughter of a forester in the Indian Colonial Service. My most recent colleague at DFID is the son of a district officer who was based for many years in Somaliland, where indeed Nick was born. Jeremy, whom I quote at the beginning of this chapter, had connections in all of the three main strands of the Empire – the commercial empire, the settler colonies and dominions and India and the colonies; a great grandfather opened up Swatou, a treaty port after the second Opium War; two of his brothers settled in Australia; and his grandfather joined the Indian Civil Service where he had a distinguished career as a lawyer. The Royal Navy carried people to posts in Ceylon; the military to Indian commands. Global organisations like Cable and Wireless meant that people operating the network could be posted anywhere from Persia to St Helena. One had a great uncle who was part of Cecil Rhodes' takeover of Central Africa through the British Central Africa Company. I suppose there is nothing mysterious in the passing of aspiration and even career from parents to children; many actors are children of actors, writers are children of writers. We respond to the world in which we grow up. In fact, among many of these international economists, the mantle has fallen to the next generation, who are working in various parts of the Commonwealth. Empire has spun a web in so many ways over my generation. Even where there was no direct involvement in the Empire, anyone who served in the Second World War would have found themselves in some part of the world defined as under the British flag; the sun never set.

So, to go back to motivation, adventure and exploration is important. I have worked on 40 or 50 projects in 30 or 40 countries – and have visited over 70. I have advised on banana marketing in the Windward Islands, village water supplies in the Sahel, heavy power plant in India and the motor industry in Pakistan. I have written environmental management plans in

China, in Syria and in Mauritius. I have worked on post-war reconstruction projects in Egypt and the Lebanon. I have sat in more airport waiting lounges than most and eaten more local cuisine than most. I have sat in ministries in ante rooms for hours at a time and have understood more about places than is ever possible as a tourist. It has been a fantastic life. But the question of what it – I – have achieved is hard to sum up. The world of aid and technical assistance is paved with good intentions but what we nowadays call 'outcomes' may not be what was planned. A few years ago the OECD tried to codify best practice in aid in fragile states – states suffering from conflict or vulnerable in some other way – and concluded that one of the most important axioms should be 'Do no harm'. Not an ambitious target perhaps, but looking at changing understanding and even fashions in aid over the past 30 years, it might be something one would be pleased to say after a career of meaning well.

This is not the place to reflect on the wider outcomes of international aid over the past 60 years with its swings and fashions in policies and mechanisms. But it is worth making a couple of points. In the colonial days, the technical specialists and administrators were required, within the terms of their employment contract, to learn the local language(s) in the places where they were working – and they might spend a lifetime in the country. This is no longer the case. At government level, aid recipients are expected to work in English or French – the two great colonial languages; or through interpreters, which must reduce mutual understanding. Secondly, the short term nature of contracts today means that people cannot develop deep knowledge of their development partners. Long-term contracts mean two or three years; and organisations like the World Bank blow into a country with a team of five or ten specialists, experts (like me), and stay for perhaps a week or two, make intense demands on the local administration, and then depart. Projects and programmes are too often determined in ministry buildings in the capital, while the needs and wishes of the people in the field are either ignored or taken for granted. So whether one can contribute much with short visits to a country where one talks through an interpreter is also a question that should be asked. Technical assistance should be a long-term process – and the 20 or more years that Owen and his colleagues spent in a single country, with fluency in the language(s) would have enabled a far greater depth of understanding and knowledge than short term technical assistance and consultancy assignments.

There was a generation between mine, the restless generation of the short-term contract, and that of my parents. These were the people who went to the colonies after the war, as the recruits of the newly formed HMOCS. I have recently seen a survey of this generation of colonial officers who are now retired and still alive – the generation of Mence, Moss, Griffin and others.[4] The survey asked them what their motivation was for joining the Colonial Service; what they thought they were doing; what were their achievements and failures; and what did the people – 'they' – think of them. Like me and the others their motivation was a mixture of service and adventure; they all saw themselves as helping to build societies that would be ready for self-government in a generation (or two or three), with a rule of law and social and physical infrastructure – schools, health services, roads and ports. Their definitions of successes were wide ranging depending on their territory and their particular interests, but on the question of failure there was much consensus: that they had to leave too soon, they had not been able to finish the job training civil servants, lawyers, police and politicians to take over the British governance structures and build a developed state. That is with the benefit of hindsight; but I wonder whether the colonial servants who were leaving Africa in the 1960s could have imagined that most of the population in these new nation states would still be living in grinding poverty, five decades later?

The fifth question – what did 'they', the subject peoples, think of you? – brought more guarded responses. Some felt there was affection, others mere politeness. Some thought that there was little affection; but almost all felt that they were respected by the native population for what they were doing and for a sense of dispassionate even-handed governance. In truth, that has been, on the whole, my feeling in my 30 years of travels in the developing world. Indeed, my direct interlocutors and counterparts in developing countries have been – more or less by definition – educated and professionals. But as a development economist one has the opportunity of meeting many types of people, exploring different business areas, poking one's nose into all sorts of situations, interviewing villagers or factory owners, and I was rarely aware of anything less than politeness and a degree of respect.

We must consider too what is meant by 'development'. Surely our underlying goal has been to strengthen countries' development processes, to contribute through training or funds or accurate assessment of the needs of a situation, but the achieving of 'development' is hard to measure, and the question of what is meant by development is an interesting one. A substantial part of my Master's degree discussed this, and there are many possible definitions. At its most basic it must be to pull the world population out of poverty so that they have a chance of living a fulfilled and enriching life, rather than be scrabbling around in the dirt of extreme deprivation, and there can be nothing wrong with this as a goal.

Other definitions of development concern the structure and state of institutions, the moving of a society where an institution – say the Ministry of Finance – is the embodiment of the minister who bestows patronage to his clients because of his institutional role; to one where the Ministry of Finance is a body of independent civil servants, whose role is to implement the policies put in place by the elected minister and his political colleagues.[5] This is a structure that we in the developed world (I hesitate as to whether to put inverted commas around this word) take for granted, and in Britain, despite a characteristic British cynicism about politicians, we are shocked to the core when it emerges that a minister has used his role to further the interests of himself, his family, his clan.

One of the subtitles for this book was 'be careful what you wish for'. It seems to me that there are many aspects of this in the brief and partial history of Kenya that I have written. It applies too to the development professional, to her shame, I think. When I first went to Kuala Lumpur (it has not only been Africa) in the 1970s, it was a small town, strongly Chinese, with low buildings and verandahs, food freshly cooked in open street stalls; a city of charm and tradition – although poverty was visible and grim. Going there in 2010 I found it a huge modern city, with skyscrapers, ring roads, mass transit, noise, and crowds; nothing of the earlier charm seemed to remain. But this is a society which has seen sustained economic growth for 15 years at more than 5 per cent per year – the city has been fully transformed into a modern one – because of growth, and because of the success of policies aiming for development – and there was little visible poverty on the streets. Of course this is right; but the old – previous – society seemed to me to have more charm. But this is a dangerous sentimentality. At least KL has the appearance of a proper modern city with few obvious pockets of great poverty. Not all cities are the same.

Singapore in the early 1970s was a delight – romantic, little back streets and a few new shopping malls, junks in the busy harbour, food markets, and flowers; but an engineer who had been there before me said in an older and wiser way 'ah but you should have seen it 15 years ago – then it was really nice'. Twenty years later I took my daughter there for the first time – and I was shocked by the height of buildings, the modernisation of the city, and when she exclaimed with her delight I responded much like the engineer. I went again to Singapore from Malaysia on the trip in 2010 – and it is once again transformed, this time to a futuristic modern city with spectacular architecture, manicured parks and gardens and fabulous food – no longer from ramshackle street markets but now in smart bars and immaculately clean food stalls. Lacking in atmosphere perhaps, but this is surely what development must aim for – not just buildings but the enrichment of the people. Old romantics may lament loss but the future must be embraced too – and it is after all why we went into the technical assistance field.

When I first went to Addis Ababa in the early 1980s, the city at night smelled of the smoke of eucalyptus fires; between the city and the airport was a band of countryside – eucalyptus plantations and fields and villages. Now the city has swollen to the airport, and is surrounded by a band of elevated ring roads. But flying over the city one sees an enormous shanty town where people live in as much squalor as they did 30 years ago. Ethiopia, despite the shiny new buildings and roads, still counts among the poorest countries of the world, and DFID still gives aid for health and education. Yet a high proportion of the country's national income is spent on fighting a relentless, intransigent war along the borders of Eritrea. Wealth it seems is being displaced away from the people of the country into a war which sustains the elite class of the society. That is surely not development.

Perhaps too – how candid can I be? – it is because 40 years ago I was of a relatively wealthy elite in global terms; now I am far from that elite, and I daresay, for all my wishes for countries to develop, that I rather regret being displaced by greater wealth held in very different societies as Britain slips into the second division in world power. But in the case of Addis Ababa and other places I feel an anger that, despite the freedom from the colonial yoke – if that is what it was – the elites have proved incapable of development in the sense of improving the quality of life for all, or of institutional maturity. Not that Ethiopia was colonised for long, a brief suzerainty by the Italians which left a legacy of a road network and the ability to make the best pasta in Africa.

I cannot speak of Nairobi in the same way since I have never worked there professionally. But in the 50 years since Independence it has become something of a by-word for corruption. Political power and patronage have alternated between Kikuyu and Luo, following Kenyatta's death in 1978. When my friend Lukas Alebe, formerly an archivist at the Anglican Church of Kenya in Nairobi and now working at Bukura Agricultural College introduced me to his director there he said 'Her father was Deputy Director of Agriculture in Kenya but *he never owned land there*'. This was not true for all administrators – Sir Philip Mitchell himself bought a farm in Kenya as did some others on retirement; but I think Lukas was suggesting that nowadays it is unusual for civil servants in the agricultural sector not to profit directly from their position in terms of land. Political power has been associated with the creation of personal wealth in a way which it was not in the tradition of the Colonial Service.[6]

We also have to consider whether our – still somewhat paternalistic – views on what poor countries should do is shared by the governments in these countries. There are those who argue that outsiders can never help a society become richer and fairer, and that all should be done within the society itself. But transfer of resources from rich to poor and of knowledge and expertise should not in itself be damaging. Over the years, the aid funds to developing countries have been vast; in 2010 alone the aid budget for DFID was £8,000,000,000; a worthy effort at transfer of resources from richer to poorer – equivalent in the currency of 1945 to £230 million. Have these transfers enabled societal transformation in the two or three generations since Independence? Perhaps, to some extent.

At the time of Independence for Kenya and her neighbours, the thinkers of the Colonial Office had seen their role of trusteeship as lasting for two or three generations but it was not to be – as nationalism grew in the colonies, as the power of America and her distaste for the European colonies dominated world polity – and perhaps as the Cold War gave cause to the Great Powers to find allies through their own forms of patronage. If our predecessors felt their contribution cut short, have we been able to make good that deficit? Hard to say. There are no counterfactuals, but in some ways it may be that the colonial servants in the early part of last century did more to establish the basics for development than the aid professions have been able to offer later. The parallels cannot be drawn too closely; in the colonial period the government was British, whereas aid flows of the past 50 years have gone to sovereign governments or projects within independent countries. And I am convinced that aid has made a positive contribution and at the very least helped to protect some of the poorest in society.

Certainly the backbone of agriculture, now the mainstay of Kenya's economy, built directly on the experimentation, and efforts of the colonial period – the officials and the settlers.

I am not trying to justify empire or the colonial experiment. I simply think it was inevitable. The best we can hope for is that the rulers were – on the whole – fair, just and honest. Not perfect, but good enough. In this book I have tried to show that this was true for a certain set of colonial servants in a certain place at a certain time.

9

EPILOGUE

When Owen and Dorothy left Kenya in 1950, they were determined to develop their own farm in England, for Owen to put the theoretical and hard physical work of the previous 20 years into practice, no longer as an adviser, but as a farmer and grower. In 1951 they bought White Hart Farm at Stocking Pelham in Hertfordshire, a white plastered four square farmhouse with a beautiful brick built barn which has since been designated Grade II listed. The farm was small, perhaps about 50–70 acres, mixed farming with some wheat and barley, pigs and chickens, an orchard and soft fruit. In the early 1950s, farm mechanisation was largely still to come. They owned a small grey Ferguson tractor – always known by everybody as Fergie – but when it came to harvest a contractor came in with the combine harvester.

My memories of it are of sheer delight; feeding worms to the chickens, 'helping' Owen roofing the chicken sheds, caring for newly hatched chicks on the back of the kitchen range in winter, watching moorhens with their jerky head movement on the farm pond. Owen was happy, but it proved hard for them. This size of farm would be marginal for a second full time man – both in terms of man-hours required and in terms of the net farm income being sufficient to pay a semi-skilled second man – but very hard work for one man, no longer in the first flush of youth.[1] It was a feasible size; incomes from farms of that size could have been as high as £1,900 per year, although many earned less.[2] Typical net incomes from a small farm were around £1,000 per year, but the small farm sector was struggling. And they were in their forties then, and Dorothy was – to say the least of it – unused to hard physical labour. Money was relatively scarce; they had their Colonial Service pension but they needed to make a living out of the farm as well.

Since her first pregnancy in Kakamega, Dorothy had suffered a recurrence of phlebitis, which, with the hard work of the farm, developed into varicose ulcers. She was perhaps weakened by her four pregnancies – and the life of an official's wife in Kenya was not the life to build physical strength and stamina – a little tennis perhaps but no great exertion. (Having said that, she lived to 93 and, apart from the varicose ulcers which never really left her, she was robust and healthy until a final stroke.) Owen's health legacy was somewhat different; he had malaria many times in Kenya and when he became ill in the UK with unexpected infections, doctors sought always to diagnose them as recurrent bouts of malaria. But in the end – after years of misdiagnosis – they were pinned down to a poisoned gall bladder and he died after surgery in 1972 at the age of 68.

After two years on the farm, Owen came to the conclusion this was not the life for them. He applied for various jobs but there were few around, he was relatively old and the Kenya and colonial experience was not valued in contemporary post-war Britain. Indeed they were rather out of kilter in this very different world and found it difficult in some ways to fit into English life. The early 1950s was still a time of rationing and post-war bleakness in England; the disconnection between the world of ration books and Dorothy's Kenya days is illustrated in a tale she ruefully told against herself: when she returned with the baby and went to stay with her parents in Reigate, she went with her mother to the butcher, and cheerfully asked for half a dozen eggs – in front of a shop full of people. Her mother was furious with her – 'I'll never be able to go there again, how could you embarrass me like that'.

It was a drab time in England, and those who had been out of the country for a long time found conditions very difficult – the queues, the shortages, the lack of heating, the poor transport. Rationing applied to clothing, food and petrol. Above all, they had not shared the war years in the same way. Owen had not fought – he would have been too old for active service although men of his age served in various roles in the conflict areas. For many in England the war was a period of emotional austerity as well as fear – husbands were away at the front, wives left at home and couples not seeing each other, babies born and reared singlehandedly by the women. In contrast, Dorothy and Owen had been leading a rich sundrenched life in Mombasa, free from rationing, although perhaps not fear.

Owen found his vocation in the church, and entered Westcott House Theological College in Cambridge in the autumn term of 1952. This was not an uncommon path for ex-colonial officers in fact. From among his contemporaries Ray Barton entered the Ministry, studying at Ridley Hall, Cambridge, from July 1955 until 1957 when he was ordained as a priest, and went on to be Rector of Nettlestead, near Maidstone, Kent, where he remained Rector until he retired in 1970. John Booth had left Egerton and returned to the field but when he retired he became a barrister, a different approach but still one of the professions open to those with the unfashionable colonial experience.

The forthright Donald Storrs-Fox also became a priest – but shocked Owen and other friends by converting to Roman Catholicism. In fact, among Donald's lively correspondence from Kenya there are many letters home[3] discussing theological matters with his customary intensity. Whether the Church was Dorothy's vocation is another and unanswerable question, although I can conjecture that it is not where she would have seen herself in middle age. Be that as it may, it turned out that Leonard Beecher's prognosis for Owen – that some village church somewhere in rural England would find itself with the services of a talented church musician – would turn out to be truer than expected, and the parish was that of St Giles Church, Totternhoe, in Bedfordshire. Owen's oratory and music, Dorothy's organising and administrative skills (president of the Women's Union, Chair of the Women's Institute – these were positions that came with the role of vicar's wife) found a path in this different type of ministry. In fact, she became popular on the talkers' circuit for local organisations like the WI – she gave many talks about life in Kenya with photographs: safari and camping, the domestic details like the canvas bucket shower under a tree by the tent, and entertained with stories of staffing matters (a favourite was how a kitchen boy, instructed by his memsahib to prop the toast up in a toast rack once cooked so that it would not go soggy, was seen the following day, sitting on the floor and toasting bread on the open fire with the fresh slices of toast propped up between his bare toes).

Owen continued to make music, and when he retired from Totternhoe in the early 1960s, the village choir gave a concert – of the same programme as the memorial concert at Mombasa Memorial Cathedral in 1944, including, memorably, Gounod's 'Gallia'. Then they bought a bungalow by the sea in Whitstable with Dorothy's legacy from her father. It is a curious thought that apart from their brief time on the farm in the early 1950s, this was the first home they had owned during their 32 years of marriage, living all the time in homes associated with the job.

Owen was not alone in finding life after Kenya difficult. He had reached the top of the career ladder in Kenya, and was now starting again, first as a farmer and then as a priest – at the bottom. He and his generation were returning to an England much changed by the war, and to an environment which did not value the work they had been doing in the colonies. When they had joined the service, it was to 'guide the African peoples through the transition from tribalism to nationalism'. In a spirit of paternalism, they had tried to improve the lot of the African, they had worked hard to encourage better crops, better livelihoods, to improve standards of husbandry and to encourage exports and new crops like rice and cotton, and later coffee. The language of the Agricultural Department reports could be patronising but there is little doubt that the overwhelming motivation of these colonial servants was to do well by the African with whatever resources they had. But in the England of the 1950s, the image of life in the colonies was coloured by African nationalistic uprising; Mau Mau attracted considerable sympathy in England, at least until the Lari massacre. But later, after the news of brutal treatment of inmates at the Hola camp, the colonial servants became tarred with a brush of oppression. The work and contribution they had made to the lives of Africans was ignored, forgotten.

NOTES AND REFERENCES

Chapter 1

1 Society for the Promotion of Christian Knowledge.
2 Margery Perham, *The Colonial Reckoning – The Reith Lectures*, Fontana, 1963.
3 Mackenzie, 1999. 'The popular culture of empire in Britain' in Judith Brown and Wm. Roger Louis, *The Oxford History of the British Empire – The twentieth century*, OUP, p218.
4 Famous to today's generations from the film *The King's Speech* of 2010.
5 Lugard, *The Dual Mandate of the British in Tropical Africa*, 1922.
6 A. Kirk-Greene, *On Crown Service*, 1999, p186, I.B.Tauris.
7 Ibid.
8 Ibid.
9 Formerly known as the South Eastern College of Agriculture.
10 There are few careers today where the salary for a senior post is only 2.5 times greater than the starting salary.
11 Kirk-Greene, op.cit.
12 M. R. Dilley, *British Policy in Kenya Colony,* Frank Cass and Co, 1965 (second edition).
13 Department of Agriculture, 1927.
14 In 1929 Kenya was divided into ten provinces, each under the supervision of a provincial commissioner who acted as liaison officer between the District Councils and the Local Native Councils in his area. In the provinces were some 870 districts, each with its own district officer. Each district had at least one agricultural officer, a police superintendent, a medical officer, sometimes a veterinary or forest officer, perhaps an officer from the Public Works Department.
15 Department of Agriculture.
16 The historical background on Meru is drawn from papers by Titus Meruthi, a local historian in Meru. While at college, he undertook a project to write about his home area – and realised there was

almost nothing written, and he has spent the past few years writing up the history of Meru since the arrival of the white man.

17 Ibid.

18 Oates, 'Reflections on changes over time', unpublished manuscripts held in Rhodes House.

19 Hill 1949, *The Permanent Way*. East African Railways & Harbours.

20 The swarm could be easily one mile wide, and at a density of perhaps to 100 or 200 million locusts per square mile, this swarm covering 600sq miles would contain at least 60,000 million locusts. The overall weight of the swarm might be 120,000 tons (at an average weight of 2gms), therefore it would consume 120,000 tons of crops and grazing each day.

21 Sunman, AGR 1928.

22 G. B. Masefield, *A History of the Colonial Agricultural Service*, OUP, 1972.

Chapter 2

1 Sir Edward Grigg, Governor of Kenya 1925–30, in a speech in Falmouth, March 1927.

2 Charles Miller, *The Lunatic Express*, Macmillan, 1971, p168.

3 Elspeth Huxley, *White Man's Country – Lord Delamere and the making of Kenya*, Chatto & Windus, 1935, p34.

4 Descriptions of slave conditions taken from Miller.

5 Hill, op.cit.

6 Hill, op.cit.

7 '*The Fashoda Incident of 1898 – encounter on the Nile*' by Darrell Bates gives a vivid account of the Fashoda Incident.

8 Huxley, op.cit. p34.

9 Hill, op.cit. p170.

10 The term 'Lunatic Express' was coined by the radical Whig MP Labouchere, an implacable opponent of the railway, following a particularly acerbic run-in with Lord Curzon, (Under Secretary of State for Foreign Affairs in the Liberal government of Lord Rosebery) and supporter of the line, in the following verse, which was published in the political magazine *Truth* in 1895 (which Labouchere himself had in fact founded):

What it will cost no words can express;
What is its object no brain can suppose;
Where it will start from no one can guess;
Where it is going to nobody knows;
What is the use of it none can conjecture;

What it will carry, none can define;
And in spite of George Curzon's superior lecture,
It clearly is naught but a lunatic line.

11 Extracts from Miller.
12 Hill, op.cit. p152.
13 *The Man Eaters of Tsavo*, Lieutentant Colonel J. H. Patterson, Macmillan, 1934. Patterson was an engineer whose task was to supervise the building of a permanent bridge over the Tsavo river.
14 This is an astonishingly low figure; for comparison, recent rehabilitation of a railway line in Angola cost about £1.5 million; a mainline railway in China in 2011 cost about £10 million (*Railway Finance Magazine*).
15 Elizabeth Watkins, *Oscar from Africa – The biography of O. E. Watkins*, Radcliffe Press, 1995, p2.
16 John Karanja, *The Role of Kikuyu Christians*, 2000.
17 Mary (Grimes) Crawford, *By the Equator's Snowy Peak: A record of medical missionary work and travel in British East Africa*, Church Missionary Society, 1913.
18 Ibid.
19 Ibid.
20 Ibid.
21 L. S. B. Leakey *Mau Mau and the Kikuyu*, Methuen and Co, 1952.
22 Mary (Grimes) Crawford, op.cit.
23 Church Missionary Society (CMS) was the pioneer; there were also the Church of Scotland Mission, the African Inland Mission and the Gospel Missionary Society.
24 LSB Leakey op.cit.
25 Sir Charles Eliot was appointed Commissioner in 1900.
26 Huxley, p78.
27 L. Winston Cone and J. F. Lipscomb (eds) *The History of Kenya Agriculture*, University Press of Africa, 1972.
28 Eliot.
29 Huxley, *White Man's Country, Lord Delamere and the Making of Kenya*, Chatto & Windus, 1925.
30 Huxley, *Out in the Midday Sun*, Penguin, 1985.
31 Some fascinating characters followed the missionaries. Mervyn Hill, chronicler of the railway in *The Permanent Way*, arrived in Kenya as a young man of around 20. Ostensibly he was sent by his family to the Highlands to farm as he was highly asthmatic and it was thought this would be good for his health. But he was a roguish young man, and much more attracted to the horses in Nairobi than to farming – he later became secretary to the East African Stud

Book in the 1930s. He married a young lady called Sunbeam Croxford whose family came from South Africa to farm in Eldoret. By the time they arrived in Kenya, the Croxford family had already moved from Orange Free State to South Africa, in search of cooler dryer climes. They then went to Victoria in British Columbia, but farming there failed and in 1915 they returned to the western hemisphere arriving in Kenya in 1916. After marriage in 1926 Mervyn and Sunbeam together made the new Show Ground in Nairobi. He planted trees while she was responsible for the flower beds and in particular the rose beds, needing to ensure they were in full flower for the main Nairobi show. Their first child was born at the Nairobi Show Ground, at three on a Saturday afternoon during a show. Sunbeam became secretary to the Jockey Club of Kenya at the racecourse in Kariakor, while Mervyn continued a more social life at the Muthaiga Club.

32 Bennet, *Kenya – A political history*, 1963.

33 Colonial governments were based on a classic separation between the legislative and executive functions, under the leadership of the incumbent governor, who was in turn appointed by the British government in Whitehall. The Executive Committee, Exco, can be roughly likened to the modern cabinet, made up of appointees from within government with key executive roles; the director of finance, the head of the public works department and heads of key governmental departments (including agriculture for example) would be appointed to Exco for the implementation of policy. Legislative Council was more equivalent to modern parliament, with a role in determining – or at least influencing – policy direction which would then be implemented through the Executive Council. Legco represented the interests of different groups within society, and is the locus where people or interests outside the government could express their views and exert their influence. The struggle over representation on Legco defined the politics of Kenya between the wars, with the settlers' leadership struggling both to expand their own representation and to limit that of other key groups – Indians and Africans. Both committees were chaired by the governor.

34 Robert G. Gregory, *South Asians in East Africa – an economic and social history*, Boulder, Colorado, USA, 1993.

35 The working currency in British East Africa was the Indian rupee until 1920 when, with the declaration of the Crown Colony, Kenya converted to sterling at the rate of Rs1 to 2/-. This was a rate which caused considerable hardship for some of the farmers and traders, since the exchange rate reflected the rapid rise in the value of the

rupee which was tied to silver, following the rise in the silver value after WW1. Before the war the exchange rate had been Rs1 to 1/4d but after the war as the value of silver rose the value of the rupee rose to 2/- and then 2/4d. Exporters who were paid in sterling received considerably fewer rupees than before, although their domestic costs were denominated in rupees.

36 Historians such as Matheka, Maxon and Ochieng' have argued that the introduction of the cash economy undermined traditional risk management strategies. For example, when there was surplus from a season's farming most was stored rather than traded, and held as security in case of crop or rain failure in the next season. Cash reserves might be less readily available as security in the face of famine, not least because they might have been spent rather than saved and, in certain areas traditionally vulnerable to famine, the transition from grain to cash caused severe problems for the native population.

Money changed other relationships within Kikuyu society too. Marriage has traditionally been strengthened by the payment of 'marriage insurance', a payment made by the man's family to the bride's family, generally in sheep and goats and sometimes cattle; the animals could not be sold until it was clear that the marriage would not break down. But, Leakey points out, 'nowadays, although the old custom of paying marriage insurance (or *ruraciol*) is still retained in form, it has completely lost its meaning... and marriage insurance payments are often made in cash. Such cash payments having been made, it is not at all easy for the bride's family to resist the temptation to spend the money forthwith, perhaps to buy things they need like clothes and food or a plough or motor car. And so, what used to be an insurance for the marriage, and could be returned in the event of a marriage breakdown, now becomes more and more a payment for the "purchase" of a wife. Furthermore, the pressure falls on the young man to save money from cash earnings to make the payment – 'but', Leakey says, 'a man may borrow the amount he needs "to hasten the wedding day" – and may find himself in debt for years to come'.

37 Piers Brendon, *The Decline and Fall of the British Empire 1781–1997,* Vintage Books, 2008, p355.

38 Ibid.

39 Huxley, *Out in the Midday Sun,* Penguin, 1985.

40 Thus Sir Charles Eliot, first Commissioner of the Protectorate, highly educated, extremely intelligent and refined, could refer to 'astonishing exhibitions of nudity' among the Masai, and confess

that 'nakedness' was 'one of those African customs which I would fain see done away with'.

41 Not everyone welcomed this vision of civilisation, and the settling of Kenya as a protectorate had of course not been totally peaceable. Various punitive expeditions were undertaken to subdue different tribes; but in fact the administration had a distaste for punitive expeditions: Eliot noted that martial exercises should not be 'allowed to override the greater claims of justice and good policy... if British objectives could be achieved by showing the flag rather than brandishing the sabre'. There were guns in use in the pacification – but that was as often between warring tribes as in controlling individual ones.

Chapter 3

1 M. Perham, *East African Journey*, Faber and Faber, 1976.

2 Although always known as the rhinoceros whip, it was in fact made from hippopotamus (kiboko is Swahili for hippopotamus).

3 Number two to the governor.

4 Huxley, 1985 op.cit.

5 He was elected president of a Convention of (settler) Associations (which was to become known as 'the settlers' parliament') in 1911.

6 One of the most fashionable London hotels of the day, run by Swiss hotelier Cesar Ritz.

7 Letter in *The Standard*, 1933.

8 There is still a May Mukle cello prize at London's Royal Academy of Music.

9 Writing in *Art Magazine* in 1952, A. J. R. Master refers to 'gifted musical families living in Limuru'.

10 Character in *The Mikado*.

11 The Railway Institute used to perform in the European Railway Club.

12 Nicholls, 2005, *Red Strangers: The White Tribes of Kenya*.

13 Reference to a ditty of the time which went:
Would you like to sin
With Elinor Glyn on a tiger skin?
Or would you rather err with her
On some other kind of fur?

14 Quoted in Piers Brendon, *The Decline and Fall of the British Empire, 1781 to 1997*, Vintage, 2008.

15 Brendon.

16 A. E. Harrar, unpublished memoirs and letters of service in Kenya, Tanganyika and Uganda, Rhodes House.

17 Huxley, *The Mottled Lizard,* Pimlico, 1962, p39.
18 Indebted to George Bennett, *Kenya – A Political History, the colonial period*, OUP, 1963, for the early political history.
19 The Duke of Devonshire was Member of Parliament for West Derbyshire from 1923 to 1938.
20 Dr Arthur served as representative of African interests on Legco from 1924–26 and from 1928–29 on the Kenyan Executive Council. Arthur was particularly concerned with problems of education, land ownership and labour reforms, and was involved in debates over the practice of female genital mutilation among the indigenous population.
21 Happily the liquidator managed to sell the land to Brooke Bond and Finlay Muir tea companies for the reasonable sum of £3.10s an acre (in 1920 land suitable for maize was selling for £3–5 per acre in the Rift) – which yielded funds to pay off all the colony's debts and make a payment of £800 to each of the surviving BEADOC members. This was the start of the tea industry in Kenya.
22 During the war, linen had been used for uniforms and tents and the burgeoning aircraft industry – at that time the wings were made of linen stretched around wooden frames. Britain had mainly imported flax, but supplies from Western Europe were limited due to German occupation, and Russian supplies were disrupted by the Revolution. By 1919 the price of flax rose to over £500 per ton. The scheme was expected to be a real winner – revenues for the government who subsidised the project; livelihoods for the brave disabled officers; and profits for the coordinator of the scheme, Lt Col. Hughes Ridge. He stated that 'Flax grown in East Africa is not a speculation. It is a certainty'.
23 The largest areas in Kenya province, Naivasha, Nyanza and Ukamba, were distributed as follows: Kenya province (Embu, Fort Hall, Meru and Nyeri) – 7.2 million acres; Naivasha – 4.0 million acres; Nyanza – 5.4 million acres; and Ukamba – 14.6 million acres.
24 Department of Agriculture, 1923.
25 Ibid.
26 *The Oxford History of the British Empire, Volume IV, The twentieth century*, Judith Brown and Wm. Roger Louis (eds), 1999. OUP.

Chapter 4

1 The 1930 AGR reports that towards the end of 1930 a Brazilian coffee scheme – the Brazil Valorisation Scheme – came to the end

of its resources, and coffee was released more freely onto global markets, causing a glut and falling coffee prices.

2 June Knowles (aka Elizabeth Watkins), pers. comm. 2012.

3 AGR 1930.

4 Although 'loans to the former will not be possible until a different form of land tenure for natives is instituted' – it is not clear what is meant by this. It may have been a wish rather than an actual policy.

5 Letter from Olga Watkins to her mother in Watkins, *Oscar from Africa,* p210.

6 H. Fearn, *An African Economy – a study of the economic development of the Nyanza province of Kenya 1903 to 1953*, OUP, 1960, p130.

7 Draws on Nicholls, *Red Strangers – the white tribes of Kenya*, Timewell Press, 2005, p203. Ref. also Kitson, Report on application by the Tanganyika Concessions Ltd company for exclusive prospecting licence in Kavirondo in Kenya,1933.

8 John L. Riddoch, *The Kakamega Gold Rush*, p143 (Ayre).

9 Various, much from Ayre's notes.

10 Nicholls, *Red Strangers*, 2005 op.cit.

11 Kenda Mutonga, *Worries of the Heart,* University of Chicago Press, 2007, p85.

12 Compensation was paid as a form of commuted rent over a period of 21 years, and the economist Hugh Fearn suggests that the compensation was very small indeed and did not allow for the improvement of land under African farming in the intervening period. Furthermore, the principal of commuted rent – which effectively removed the farmer's rights to his land – may not have been fully comprehended by the peasant farmer, whose understanding would have been based on customary practice whereby he would allow another African (usually a kinsman) the right to use his land for a period of time, but secure in the knowledge that the land would be returned to him at any time he himself wanted to use it. This was clearly not the meaning of the compensation as granted by the government.

13 AGR 1928, p4.

14 A measure of this was that five months' service counted as six for colonial servants, so that leave was more frequent than for those posted in more conducive situations.

15 John Howard, Administrator, letters to his mother in Rhodes House Library.

16 *Kenya Settlers Cookery Book and Household Guide*, published East African Standard, 1932.

17 Bill Barton, *In and Out of Africa – memoirs of a medical officer in Kenya*, 2003.
18 Mary Hamilton, *Turn the Hour,* The Book Guild, 1991.
19 Michael Blundell, *A Love Affair with the Sun,* Kenway Publications, Nairobi, 1994.
20 AGR 1931.
21 AGR 1934.
22 Fearn, p166.
23 Owen, 1990, *Leonard and Gladys Beecher – A life of service to God in East Africa.*
24 1948 Beecher Report on Education.
25 Anne Thurston, *Smallholder Agriculture in Colonial Kenya: the official mind and the Swynnerton Plan*, 1987, p7.
26 Thurston, op.cit.
27 Liversage, Unpublished memoir, Rhodes House.
28 AGR 1928.
29 Letter from Owen to Wye College Alumni magazine, 1935.
30 Ismaili.net – Heritage F.I.E.L.D.
31 Roughly £4,000 in today's prices.
32 W. Kelly, Diaries and memoirs of service. Bodleian Library/Rhodes House, 1936 onwards.
33 Letter dated 15 April 1937, Kisumu.
34 Ibid.

Chapter 5

1 Storrs-Fox letters, Rhodes House.
2 Ibid.
3 From Wye college magazine, *Agricola*, 1936.
4 Hughes Rice manuscript, 1979.
5 Ibid.
6 Ibid.
7 Pers. comm., discussions at Egerton University, 2010.
8 B. F. MacDonald, *East Africa in Wartime, memoirs of a bank manager*, held in Rhodes House. He wrote entertainingly of the secrecy around invasion of Madagascar.
9 Mervyn Hill, p534.
10 M Hamilton, 1991, *Turn the Hour*, The Book Guild.
11 C. Nicholls, *Red Strangers – the white tribes of Kenya*, Timewell Press, 2005, p216.
12 Sutherland and Cantwell, *Air War in East Africa*, Pen and Sword Publishers, 2009, p20.

13 Sutherland and Cantwell.
14 Winston Churchill, *The Second World War, Volume II, Their finest hour*, The Reprint Society, 1949, p338.
15 Ibid.
16 Churchill – 'Alone, Egypt and the Middle East', in *The Second World War, Volume II, Their finest hour*, The Reprint Society, 1949, p340.
17 The force included Somaliland Camel Corps, some of the King's African Rifles from Nyasaland, the 1st Battalion Northern Rhodesia Regiment, the 3rd Battalion and the 2nd Battalion Black Watch.
18 The capital of British Somaliland was moved to Hargeisa in 1941.
19 Apparently 'in order to test out the Coastal Defences, the Civil Defence Service and the Anti-invasion schemes' the public was told that, without further warning, their whole economic life would be disturbed and disrupted for a short time, and all naval, military and air force units would be involved. A State of Emergency would be declared which would last for several days. Very considerable preparations were made for the test evacuation of Mombasa; and eventually the call to action came:

> not by means of martial music but by the bell of the Nairobi fire engine which toured the town with an excited escort of small African boys. The streets were cleared, banks and shops closed and people took up posts at air raid post (ARP) centres and anti-aircraft posts.... The exercise involved acting out the responses to an invasion; 'bombs' were thrown, 'casualties' were brought to ARP posts, air raids by the 'enemy' took place on Nairobi. The Kenya Defence Force (KDF) stood on guard at anti-aircraft posts and gun posts surrounding Nairobi, and 'hyenas laughed under an African moon'.

When the exercise was over, and East Africans returned to work, they realised that all these activities had been an elaborate cover for real military and naval movements of the utmost importance.

> Whilst landings of invading forces were being practised up and down the coast, and troops and ships were being rushed out to repel such attacks, the Navy had slipped quietly out of Mombasa with a large convoy of transports and, a few days later, the real attack on the Madagascar ports had been successfully made!

20 Michael Smith, *The Emperor's Codes*, Biteback Publishing, 2010.
21 Pers. comm., Susan Wood, Luke's daughter, 2010.

22 Lipscomb, who chaired the production board, boasted in his book on African agriculture that 'the Increased Production of Crops Ordnance, a system which was refined and polished over later years, was commended by the World Bank mission, 20 years later, as 'a very sophisticated and efficient organisation'.
23 Cone and Lipscomb, p85.
24 In 1942 the Kenya government passed a Defence Regulation whereby compulsory labour was allowed on European farms as an essential undertaking for the war effort.
25 Letter in Rhodes House from Betty Storrs-Fox.
26 Oates, 'Reflections on changes over time', unpublished manuscripts held in Rhodes House.
27 Betty Storrs-Fox, op.cit.
28 Letters, RH, dated Kisii December 1941.
29 From a brass plaque on the wall of the vestry.
30 Gallia, *Solitary lieth the city*, C H Gounod.
31 Extracts, Owen's diary 1944.
32 £20 2010 values.
33 Extracts, Owen's diary 1944.
34 Watkins, 2005, Olga in Kenya.
35 Denzil Blunt, Department of Agriculture, 1938.
36 Ibid.
37 Mervyn Hill, *Kenya Weekly News*, 1949.
38 The Agrarian Problem.
39 Bennett, *Kenya – A Political History, the colonial period*, OUP, 1963.
40 Bennett, op.cit. p95.
41 Bennett, op.cit. p98.

Chapter 6

1 Howard Johnson, 1999. In *The Oxford History of the British Empire*, Judith Brown and Wm. Roger Louis (eds) *The twentieth century*, OUP, p613.
2 To put this in context, in 1945 the total government budget of Kenya was £6.8 million and the budget for agriculture (plus forestry and veterinary services) was £529,762.
3 E. Huxley, *A New Earth – an experiment in Colonialism*, Chatto & Windus, 1960, p188.
4 Mackenzie, a historian reviewing agriculture in Kenya in 2000 in a largely unsympathetic light notes that, between 1930 and 1938,

the advice given by agricultural officers tended to be 'congruent with African farming methods'.

5 Colin Maher, Memorandum on dealing with soil erosion in Machakos, 1937.
6 Liversage, op.cit.
7 Department of Agriculture, 1947, p17.
8 Blundell, *A Love Affair with the Sun*, 1994, p27.
9 Hennings, p95 in Cone and Lipscomb.
10 Figures are difficult to interpret; the Blue Book totals are much higher than those shown in Agricultural Department reports up to 1935 – it is not clear why.
11 Francis Loyd, private secretary to the governor in 1942 and later DC in Kakamega.
12 Huxley, op.cit.
13 Swann, 1982, ODRP papers.
14 Report of agricultural officer for Kiambu, 1945.
15 Clayton, quoted in Cone and Lipscomb.
16 Huxley, p191.
17 When Philip Bowcock, a DO in the Sudan, visited Kenya in 1961 he was struck by the rich lands and agriculture in the country; specifically 'Machakos, beautifully terraced with tall waving grass'.
18 Hennings, memoirs recorded for ODRP in 1979, and others.
19 Beecher, 1949. Native population in 1945 estimated at 3.9 million; assume 40 per cent under 20, 20 per cent under 10, then under 10 population is 780,000; so perhaps there was only one in three to four children in education, and mainly boys. Education was also expensive.
20 Negley Farson, *Last Chance in Africa*, Victor Gollancz, 1953.
21 Dick Cashmore in John Johnson, *Kenya, the Land and the People*, John Johnson, 2009.
22 From Richard Frost in *Race Against Time*, quoted in Veronica Owen's compilation *Leonard and Gladys Beecher – A life in service to God and East Africa*, 1990.
23 Swann, op.cit.
24 B. Berman and John Lonsdale, *Unhappy Valley: Conflict in Kenya and Africa*, East African Educational Publishers, 1992.
25 Berman, 'Bureaucracy and incumbent violence' in *Unhappy Valley Book Two: Violence and ethnicity*, Lames Currey, 1992, p235.
26 Berman, op.cit.
27 Kenyatta, 1945.
28 Swann, op.cit.
29 James Foster, pers. comm. 2010.

30 Tony Swann, DC Nyeri, also comments by Francis Loyd, ODRP, 1979.
31 Huxley, *Out in the Midday Sun,* p88.
32 Francis Loyd, later DC Nyeri.
33 Moss, *Quavers near the Equator*, 2010.
34 A part of the army whose duties are light engineering.
35 From a note for a completion fund appeal, 1950.
36 Evan Hopkins, *A Slightly Venturesome Life, a collection of memories*, 2001.
37 Evan Hopkins, op.cit.
38 He suggests that the lack of Africans in the cathedral is because they did not speak English; certainly there had been many services by then where African priests participated, such as the enthronement of Bishop Crabbe in 1937 where the procession included six African clergy together with Indian Christians.
39 Hyslop, 'The prospects for music education in Kenya', government report, 1964.
40 AGR 1946.
41 The British Council had been established in 1934 to promote British culture and language.
42 BOAC began a service with the Solent flying boats from Portsmouth to South Africa.
43 Leonard Beecher, *East African Standard*, 14 October 1949.

Chapter 7

1 Hubbard, 2010, *The United States and the End of British Colonial Rule in Africa, 1941–68.*
2 This section draws on Hyam's excellent summary of policy under the 1945 Labour government, *Understanding the British Empire.* 2010. pp232–235.
3 Ibid.
4 Speech to South African Parliament, 3 February 1960. As Hyam op.cit. noted (p353) Macmillan was also giving an unequivocal warning to the apartheid government when he declared that there were 'aspects of South African policy which make it impossible to support you without being false to our own deep convictions about the political destinies of free men, to which in our territories we are trying to give effect'.
5 Hyam, op.cit. p227.
6 Hyam, op.cit. p216.

7 Thus, recruitment increased from 325 in 1939 to 1,715 in 1946 and 1,510 in 1950, declining after that to 983 by 1957. The total number of overseas posts increased from 8,000 in 1938 to 11,000 in 1947, and a much higher proportion were technical posts – health and education for example – than at earlier times. By 1960, the HMOCS numbered 20,500 officers, a higher number than the Colonial Office had ever had.
8 East African Annual, 1944–45.
9 East African Standard publications.
10 Mitchell, 1947, *The Agrarian Problem.*
11 Francis Loyd.
12 Matheka, *Colonial Capitalism, Ecology and Food Crises in Machakos District 1895–63*, PhD thesis, 1993.
13 Paradoxically, much of the progress was possible as a result of the 'villageisation' of people during the Emergency; and because the leaders of the Mau Mau, who had been attacking land reform and consolidation had been interned, leaving the field free for more radical changes in agricultural policy.
14 Blundell, *A Love Affair with the Sun*, 1994, p149.
15 Roger Swynnerton in unpublished interview with Masefield, Rhodes House.
16 Dick Moss, *Quavers near the Equator,* 2010, p60.
17 These included Kamusinga Friends, Alliance Boys, Narok Secondary, St Andrews, Turi, Kenton College, Hospital Hill, Sterehe Boys' Centre and Nyeri High.
18 Sadly Roy died in late 2012.
19 Together with Tony, her husband, Valerie is a wonderful source of stories and information and powerhouse of contemporary Nairobi music making.
20 Washington Omundi had studied under David Wilcox at Edinburgh University after Hyslop had helped him to obtain a place.
21 Kofsky, 1963, quoted by Hyslop.
22 Luvai, pers. comm.
23 Hyslop, 1964, *The prospects of music in education in Kenya.*
24 In fact the organists at the cathedral in Mombasa carried on being whites well into the 1960s. Marjorie Hawkins who sang soprano at the Centenary celebrations was still the organist in the 1950s and early 1960s (and played in Nairobi). Later, in 1979 Paul Hill, who was Director of the British Council, played and taught the organ.
25 Swynnerton.
26 Dick Moss, personal communication, 2012.

27 Department of Agriculture.
28 Lawrance, *Land Consolidation and Registration in Kenya*, conference paper, 1970, p453.
29 H. Fliedner, *Land Reform in Kenya*, IFO-Inst., 1966.
30 This study has been an amazing source for me in this writing and many of the more lively comments on agricultural work have come from these interviews.
31 A. Thurston, *Smallholder Agriculture in Colonial Kenya,* 1987.
32 Thurston, op.cit.
33 Huxley, 1960, op.cit.
34 Among the new cash crops introduced in the 1950s – tea offered very good potential but getting it going proved a bold and challenging adventure. The agricultural advisers saw that the Kipsigis land, similar to the tea estates owned by Brooke Bond and James Finlay, the giants of tea in the highlands, could become profitable and enriching for the smallholder farmer. It was a challenge – tea is normally grown in large estates: how could it develop as a peasant scale crop, grown in little blocks, scattered about the place, not centralised? Getting it to the factory quickly, as the quality deteriorates rapidly (rather like *khat*), also posed a serious problem. Huxley regarded the tea project as one of the 'boldest and most ambitious ever taken on in Africa', a gamble with human nature. She asks: 'Can you turn an indolent, cattle loving, illiterate, easy going and suspicious tribesman into an industrious, steady going, intelligent and meticulous tea planter, all in a few years?'
35 Kandagor *Rethinking British Rule and 'Native' Economies in Kenya*, 2010.
36 M Macial, 1985, *Bwani Karani*, Merlin Books.
37 Mackenzie 'Contested Ground: colonial narratives and the Kenyan environment, 1920–45', 2000.
38 Alex Holm, Director of Agriculture, 1926.
39 At the time of the Emergency Kenyatta was arrested with other Mau Mau leaders and he was imprisoned for seven years in Lodwar, in Turkana, up near the Ethiopian border, in an area which was not part of the Mau Mau sphere of influence. Whitehouse ('Wouse') had been appointed there as ADC in 1946, and hence in due course became 'Jomo's jailer'. Kenyatta's time in prison there is documented in various places, and during his period there he changed from a rebel leader, 'a self seeking and irresponsible agitator, over fond of the bottle, paunchy, over given to rhetoric. A sick man with short life expectancy'. But, she writes 'he came out

of prison with his health much improved, a mature, hardworking elder statesman with a breadth of understanding rare in African politics, and a willingness to work with the British to achieve peaceful independence'. Watkins suggests that this was partly the influence of Wouse, who would talk to Kenyatta and provide him with books, and ensured his recovery and safety following an attempt to assassinate him by his fellow prisoners. She also suggests that a visit by Bishop Kariuki in 1957, who had stood up against Mau Mau, and who was engaged in rehabilitation of Mau Mau detainees, had an important influence on him. In any event, when Kenyatta was released from prison in Lodwar he said in an interview: 'We have been in a university. We learned more about politics there than we learned outside.' This did not assuage the anxieties of the white settler population in the run up to independence, but in Nakuru in 1963, to a very anxious audience of white farmers, Kenyatta gave his famous 'forgive and forget speech', see page 220.

40 Three meetings were held at Lancaster House in London to hammer out a constitutional framework and plans for independence, in 1960, 1962 and 1963. It was in the framework of these conferences that the land redistribution scheme was initiated.

41 P. D. Adams, Deputy Director of Settlement, 'Kenya's land resettlement story', 1968.

42 Adams, op.cit.

43 Op.cit.

44 Adams, op.cit.

45 Adams, op.cit.

46 Chinua Achebe 'The Education of a British Protected Child', essay delivered as the Ashby Lecture at Cambridge University. Coincidentally, Eric Ashby had been Master of Clare College from 1959–67; he was also a very close friend of my father's and it was he who gave me away *in loco patris* at my wedding.

47 Nowadays, besides the National Park, and successful coffee growing, Meru is best known as the most important area in Kenya for growing *khat*, a leafy stimulant that is known locally as *meera*. An appetite suppressing stimulant if used in moderate quantities it can be addictive and, according to Asenath, Meru now has a very high level of crime and rape, due in large part to the use of the drug. The leaf deteriorates rapidly, and the time from picking to purchase and chewing should be very fast, so the trade has a frenetic quality with trucks hurtling from field to market to ensure the freshness of the product. While much of the *meera* is rushed to

Wilson airport for export to Somalia a lot is consumed locally, and washed down with a local home brew, making the drug quite lethal, encouraging rape and violent behaviour. It enthrals young men – and even children, who are used to climb trees to pick the leaves which they can do because they are small and won't break the fragile branches of the tree. Not the sort of crop development envisaged by our young advisers from Wye College.

48 Kenda Mutongi, *Worries of the Heart: widows, family and community in Kenya*, 2007 (the research was carried out in the 1990s).

49 Ibid.

Chapter 8

1 Kirk-Greene, p75.

2 Op.cit. p215.

3 International Consulting Economists' Association.

4 Kirk-Greene, unpublished survey, lodged at Rhodes House.

5 Durkheim formalised this distinction in the terms *gemeinschaft* – a society when patronage is important, relationships are personal – and *gesellschaft* – a society of the modern industrialised type characterised by impersonal relations between state institutions and individuals.

6 Michaela Wrong – a powerful writer and scourge of the abuses of empire – articulates this in her book about corruption in Kenya politics, *It's Our Turn to Eat*, 2009.

Chapter 9

1 Thanks to Jim Reed, formerly UK Ministry of Agriculture for this insight. He adds that it would have depended also on the degree of mechanisation – there were still many heavy horses in harness in the fifties. A farm using horse power for field work required a lot more manpower than one that was mechanised – which is why mechanisation proceeded so relatively fast once the 1947 Agriculture Act had introduced deficiency payments and farm incomes rose in most years through to the mid-sixties.

2 Jones, *British Agriculture Since 1950*, 1970, conference paper.

3 Lodged at Rhodes House.

BIBLIOGRAPHY

Abrams, P. D. 1979. *Kenya's Land Resettlement Story, how 66,000 African families were settled on 1325 large scale European owned farms*. Nairobi: Challenge.

Achebe, C. 2009. The education of a British-protected child. In *The Education of a British-Protected Child*. London: Penguin Modern Classics.

Adams, P. D. 1968. *Kenya's Land, the Resettlement Story*. Nairobi: Government of Kenya.

Agricola, alumni magazine of Wye College. Magazines 1934 to 1945.

Agriculture, Department of (AGR) 1922 to 1950. Annual reports. Government of Kenya.

Agriculture, Department of (AGR) 1938. Agricultural census – European areas. Government of Kenya.

Agriculture, Department of (AGR) 1945, 1946, 1947, 1948, 1949, 1950. Annual report. Government of Kenya.

Anderson, D. and D. Throup. 1985. Africans and agricultural production in colonial Kenya: The myth of the war as a watershed. *Journal of African History*, 26.

Anderson, T. F. Reminiscences (of his time as DC in Kenya). Bodleian Library/Rhodes House, MSS.Sfr.s.1653.

All Saints Cathedral, 1959. *History of All Saints Cathedral*. Nairobi: All Saints Cathedral.

Baron, W. D. A. 1949–50. Safari Reports in Kenya by the DC. Bodleian Library/Rhodes House, MSS.Afr.s.852.

Barton, B. 2000. *In and Out of Africa*. Bedale, UK: Blaisdon Publishing.

Beecher, L. 1949. *African Education in Kenya* – The report of a committee appointed to inquire into the scope, content and methods of African education, its administration and finance, and to make recommendations (The Beecher Report). Nairobi: Government Printer.

Bennett, G. 1963. *Kenya – A Political History. The colonial period*. London: Oxford University Press.

Berman, B. and J. Lonsdale. 1992. *Unhappy Valley: Conflict in Kenya and Africa. Book One: State and Class.* Nairobi: East African Educational Publishers; Athens, US: Ohio University Press.

Berman, B. and J. Lonsdale. 1992. *Unhappy Valley: Conflict in Kenya and Africa. Book Two: Violence and Ethnicity*. Oxford, Nairobi: James Currey.

Black, J. 2007. The Curse of History. *History Today* Volume 57, Issue 8.

Blundell, M. 1964. *So Rough a Wind.* London: Weidenfeld and Nicholson.

Blundell, M. 1994. *A Love Affair with the Sun.* Nairobi: Kenway Publications.

Bolton, H. L. 1970. *A Life in Africa*. Privately published.

Booth, J. E. P. 1948. Egerton School of Agriculture annual report. Bodleian Library/Rhodes House, MSS.Afr.s.847.

Booth, J. E. P. 1979. Letters. Bodleian Library/Rhodes House, MSS. Afr.s.1717/17.

Cable, V. 1969. The Asians of Kenya. *African Affairs,* 68, 23.

Cantwell, J. S. and D. 2009. *Air War in East Africa 1940–41*. Barnsley, UK: Pen and Sword Aviation.

Cary, J. 1941. *A House of Children.* New York: New Dimensions Books.

Cass, Bridget, Memoirs of Hill and Hughes families. (Unpublished)

Chapman, G. R. 1971. Edgar Bryson, Missionary to Kenya. (Unpublished)

Clayton, E. S. 1961. Economic and technical optima in peasant agriculture. In *Readings in the Applied Economics of Africa,* ed. J. I. Currie and Edith H. Wetham. Cambridge: Cambridge University Press.

Clayton, E. S. 1977. Correspondence and comments on 'Agricultural Development in Kenya' by Smith. Bodleian Library/Rhodes House, MSS.Afr.s.1717/27.

Clive, J. H. 1980. A cure for insomnia 1920 to 1047. (Unpublished memoir)

Colchester, T. 1977. Beginnings of coffee growing in Embu District, and land ownership in Mwe-Tebere in the thirties. Bodleian Library/ Rhodes House, MSS.Afr.s.1717/30.

Colonial Office. 1927 onwards. *The Official Gazette of the Colony and Protectorate of Kenya*. Nairobi: Government Printer.

Cone, J. F. and L. W. Lipscomb. 1972. *The History of Kenya Agriculture*. Nairobi: University Press of Africa.

Corfield, F. D. 1960. *The Origins and Growth of Mau Mau – An historical survey.* Colony and Protectorate of Kenya. Nairobi: Government of Kenya.

Coryndon Collection. 1924. Papers and correspondence mainly dealing with land. Bodleian Library/Rhodes House, MSS.Afr.s.633,17/2.

Crawford, E. M. 1913. *By the Equator's Snowy Peak: A Record of Medical Missionary Work and Travel in British East Africa*. London: Church Missionary Society.

Dilley, M. R. 1966. *British Policy in Kenya*. London: Frank Cass & Co.

Dobbs, C. M. 1930. Administration in Kisumu, 1908 to 1930. Bodleian Library/Rhodes House, MSS.Afr.s.665/1.

Dobbs, M. W. 1950. Recollections of Kenya 1900–1931. Bodleian Library/Rhodes House, MSS.Afr.s.504.

Duber, C. J. 1992. The British East Africa Disabled Officers' Colony (BEADOC) and the White Frontier in Kenya. *Agricultural History Review*, 40.

East African Women's League. 1987. East African Women's League 1917–1987: Seventy years. EAWL Information Leaflet.

Education, Ministry of. 2005. *History and Government, Form 3 Students' Book*. Kenya: Ministry of Education.

Etherington, D. M. 1967. Projected changes in urban and rural population in Kenya and the implications for development policy. In *Readings in the Applied Economics of Africa*, ed. E. H. Wetham and J. I. Currie. Cambridge: Cambridge University Press.

Everard, C. 1996. *The Guardian Angel: A Voice from the Wilderness*. London: Minerva Press.

Fazan, S. H. 1930 onwards. His account of Mau Mau plus detailed meeting minutes/reports on relations with Local National Councils. Bodleian Library/Rhodes House, MSS.Afr.s.1579.

Fearn, H. 1960. *An African Economy – A Study of the Economic Development of the Nyanza Province of Kenya 1903 to 1953*. London, Nairobi: Oxford University Press.

Fox, J. 1982. *White Mischief*. New York: Random House.

Fuhrer, H. 1996. *The Story of Official Development Assistance: History of the Development Assistance Committee and the Development Cooperation Directorate in Dates, Names and Figures*. Paris: OECD.

Gardner, L. A. 2009. *Colonial Origins of Government Corruption? Evidence from tax collection in Kenya and Zambia*. Oxford: Jesus College.

Grannum, R. C. 1929. Report of the committee on locust menace crop insurance. In D. Sunderland, *Economic Development of Africa 1880–1939*. London: Pickering & Chatto.

Gregory, R. C. 1993. *South Asians in East Africa – An Economic and Social History 1890–1960*. Colorado, US: Westview Press.

Hamilton, M. H. 1991. *Turn the Hour*. Sussex, UK: The Book Guild Ltd.

Harrar, A. E. 1914–36. Memories covering twenty two years of Kenya, Uganda and Tanganyika territory. Bodleian Library/Rhodes House, MSS.Afr.s.1145.

Hennings, R. O. 1951. *African Morning*. London: Chatto & Windus.

Hennings, R. O. 1979. *The Development of African Agriculture in Kenya 1935 to 1962*. Bodleian Library/Rhodes House, MSS. Afr.s.1717/76.

Hill, M. 1949. *Permanent Way: The Story of the Kenya and Uganda Railway Volume 1*. Nairobi: East African Literature Bureau.

Holm, A. 1926. Review of Agriculture in Kenya. In *Agricultural, Cotton, Entomological and Mycological Conference,* ed. D. Sunderland. Nairobi: East African Standard.

Hopkins, H. E. 2001. A Slightly Venturesome Life, a collection of memories. (Unpublished)

Howard, J. 1939. Letters to his mother. Bodleian Library/Rhodes House, Micro.Afr.606.

Howard, J. 1939. Various papers: Expenses claims 1953 to 56, Machakos District Development Plan Annual reports for Mandera 1945 and 1944. Bodleian Library/Rhodes House, MSS.Afr.s.787.

Howard, J. 1950. Mountaineering in Kenya. Bodleian Library/Rhodes House, MSS.Afr.s.606/5.

Howard, M. and Wm. Roger Louis. 1999. *The Oxford History of the Twentieth Century*. Oxford: Oxford University Press.

Hubbard, J. P. 2010. *The United States and the End of British Colonial Rule in Africa, 1941–68*. Jefferson, North Carolina, US: McFarland.

Hughes, A. J. 1963. *East Africa: Kenya, Tanzania, Uganda*. London: Penguin Books Ltd.

Hughes Rice, T. 1979. Manuscripts. Bodleian Library/Rhodes House, MSS.Afr.s.1717(84).

Hutchinson, T. H. 2008. BEADOC and Beyond. Kenya. (Unpublished)

Huxley, E. 1949. *The Sorcerer's Apprentice – A Journey Through East Africa*. London: Chatto & Windus.

Huxley, E. 1953. *White Man's Country: Lord Delamere and the Making of Kenya*. London: Chatto & Windus.

Huxley, E. 1960. *A New Earth – An Experiment in Colonialism*. London: Chatto & Windus.

Huxley, E. 1962. *The Mottled Lizard*. London: Pimlico.

Huxley, E. 1985. *Out in the Midday Sun*. London: Chatto & Windus.

Huxley, E. 1987. *The Flame Trees of Thika*. New York: Grove.

Huxley, E. 1991. *Nine Faces of Kenya, Portrait of a Nation*. New York: Viking.

Huxley, E. and M. Perham. 1944. *Race and Politics in Kenya*. London: Faber & Faber Ltd.

Hyam, R. 2010. *Understanding the British Empire*. Cambridge: Cambridge University Press.

Hyslop, G. 1964. *The Prospects for Music Education in Kenya*. ed. Inspectorate of Music and Drama. Kenya: Government of Kenya.

Hyslop, G. 1964. *Since Singing is So Good a Thing – Handbook for Music Teachers and Choir Masters*. Oxford: Oxford University Press.

Hyslop, G. 1975. *Musical Instruments of East Africa*. Africa: Nelson.

Information, Department of. 1924. *Sheep in Kenya*. Development in Africa. Kenya: Department of Information.

Jeffries, C. 1956. *The Colonial Office*. London: George Allen.

Johnson, Howard. 1999. In *The Oxford History of the British Empire – The Twentieth Century*, Judith Brown and Wm. Roger Louis (eds). Oxford: Oxford University Press, p613.

Johnson, J. 2009. *Kenya – The Land and the People: District Administration 1950–1963*. Amersham, UK: John Johnson.

Johnson, M. B. 1924 to 2002. Memoir from your 'colonial' daughter. (Unpublished memoir)

Jones, W. E. 1970. British agriculture since 1950. In *Change in Agriculture,* ed. A. H. Bunting. University of Reading. London: Duckworth.

Kandagor, D. R. 2010. *Rethinking British rule and 'Native' Economies in Kenya: Tugen Economic Transformation, 1895–1963*. Kenya: Egerton University.

Kelly, W. F. P. 1936 onwards. Diaries and memoirs of service. Bodleian Library/Rhodes House, MSS.Afr.s.2229.

Kennedy, J. 1940. Life was seldom dull – the experiences of a woman in equatorial Africa. (Unpublished diary) Bodleian Library/Rhodes House, MSS.Afr.s.514.

Kenyatta, J. 1965. *Facing Mount Kenya – The Tribal Life of the Gikuyu*. New York: Vintage Books (Random House).

Kirk-Greene, A. 1980. The Thin White Line: The size of the Colonial Office in Africa. *African Affairs*, 79.

Kirk-Greene, A. 1999. *On Crown Service: A History of HM Colonial and Overseas Civil Services 1837–1997*. London: I.B.Tauris & Co.

Knowles, O. 2008. *Back Seat Driver*. Brighton, UK: Pen Press Publishers Ltd.

Lambert, M. L. History of Barengo Region. Bodleian Library/Rhodes House, MSS.Afr.s.1388.

Lambert, R. T. up to 1950. Official reports. Bodleian Library/Rhodes House, MSS.Afr.s.397.

Lawrance, J. C. D. 1970. Land Consolidation and registration in Kenya. In *Change in Agriculture,* ed. A. H. Bunting. University of Reading. London: Duckworth.

Leakey, C. H. and C. Richard. Papers. Bodleian Library/Rhodes House, MSS.Afr.s.633.

Leakey, L. S. B. 1952. *Mau Mau and the Kikuyu.* London: Methuen & Co.

Le Breton, P. 1980. *Grim's Tale.* OSPA.

Leckie, W. G. 1932. The growing of wattle and production of wattle bark in Kenya. In D. Sunderland, *Economic Development of Africa 1880–1939.* London: Pickering & Chatto.

Lipscomb, J. F. 1955. *White Africans.* London: Faber & Faber Ltd.

Liversage, V. 1963. Economic Management in Kenya. Bodleian Library/ Rhodes House, MSS.Afr.s.510.

Lonsdale, J. 1999. East Africa. In *The Oxford History of the British Empire – The Twentieth Century.* eds J. M. Brown and Wm. Roger Louis. Oxford: Oxford University Press.

Loyd, F. 1979. Manuscripts in response to ODRP. Bodleian Library/ Rhodes House, MSS.Afr.s.1717/99.

Lyttleton, O. East African Royal Commission. 1955. Report (The Lyttleton Report). London: HMSO.

Maathai, W. 2008. *Unbowed – One Woman's Story.* London: Arrow Books.

MacDonald, B. F. 1944. East Africa in Wartime: Memoirs of a bank manager. Bodleian Library/Rhodes House, MSS.Afr.s.1145.

Mackenzie, A. F. D. 2000. Contested ground: Colonial narratives and the Kenyan environment, 1920–45. *Journal of Southern African Studies,* 26.

Mackenzie, J. W. 1999, 2010. The popular culture of Empire in Britain. In *The Oxford History of the British Empire, Volume IV: The Twentieth Century,* eds J. M. Brown and Wm. Roger Louis. Oxford: Oxford University Press.

Maher, C. 1935, February. A partial survey of the agricultural resources and potentialities of the West Suk reserve, together with notes on the general state of preservation, agriculturally and pastorally, of the area. Bodleian Library/Rhodes House, MSS.Afr.s.740.

Maher, C. 1937. Soil erosion and land utilisation in the Ukamba reserve (Machakos). 40. Bodleian Library/Rhodes House, MSS.Afr.s.755.

Maher, C. up to 1950. Various papers on soil conservation. Bodleian Library/Rhodes House, MSS.Afr.s.1741.

Maina, E. A., W. A. Oboka and J. Makong'o. 2004. *History and Government, Form 3*. Nairobi: East African Educational Publishers Ltd.

Maina, J. W. and J. D. MacArthur. 1970. Land settlement in Kenya. In *Change in Agriculture*, ed. A. H. Bunting. University of Reading. London: Duckworth.

Masefield, G. 1970. Transcription of an interview by Masefield of Swynnerton. Bodleian Library/Rhodes House, MSS.Afr.s.1426.

Masefield, G. B. 1972. *A History of the Colonial Agricultural Service*. Oxford: Clarendon Press.

Mason, T. 2008. *Anthony's Odyssey*. Privately published by author.

Matheka, R. M. 1992. *Colonial Capitalism, Ecology and Food Crises in Machakos District, 1895–1963*. Nairobi: Kenyatta University.

Maule, A. 2004. *Theatre Near the Equator – The Donovan Maule Story*. Nairobi: Kenway Publications.

McHardy, W. 1938. Diary of a Journey through East Africa. In Bodleian Library/Rhodes House. MSS.Afr. r.76.

Melville, A. R. 1970. The development of coffee production by African farmers in Kenya. In *Change in Agriculture*, ed. A. H. Bunting. University of Reading. London: Duckworth.

Mence, P. 2012. Personal communication.

Mercer, A. M. 1970. Ol Kalou Salient Project, Kenya. In *Change in Agriculture*, ed. A. H. Bunting. University of Reading. London: Duckworth.

Miller, C. 1971. *The Lunatic Express*. London: Penguin Books.

Mitchell, P. 1947. *The Agrarian Problem in Kenya*. Nairobi: Government Publication.

Mitullah, W. 2010. *The Case of Nairobi, Kenya – A Study of Slums*. UN Habitat Report. Nairobi: University of Kenya.

Moss, R. 2010. *Quavers Near the Equator: Recollections of the Nairobi Orchestra and Some Associated Ensembles*. Nairobi: Richard Moss.

Murray, N. U. 1982. Archdeacon W. E. Owen: Missionary as propagandist. *International Journal of African Historical Studies*, 15, 4, 15, 19.

Muruthi, T. 2011. History of Meru. (Unpublished manuscript)

Mutongi, K. 2007. *Worries of the Heart: Widows, Family and Community in Kenya*. Chicago, US: University of Chicago Press.

National Bureau of Statistics, Kenya. 2009. *Economic Survey*. Nairobi: The Government Press.

National Bureau of Statistics, Kenya. 2009. *Statistical Abstract*. Nairobi: The Government Press.

Ngugi wa Thiong'o. 1964. *Weep Not, Child*, London: Heinemann African Writers Series.

Nicholls, C. 2005. *Red Strangers – The White Tribe of Kenya*. London: Timewell Press.

Nightingale, G. 1914 to 1960s. Memoirs. Bodleian Library/Rhodes House, MSS.Afr.s.195.

Oates, C. O. T. 1950 and later. Papers on microfiche. Bodleian Library/ Rhodes House, Micro 606/5–607/1.

Ochieng', W. R. 1989. *A Modern History of Kenya*. ed. M. Sharman, 259. London: Evans Brothers Ltd.

Ochieng', W. R. and R. M. Maxon. 1992. *An Economic History of Kenya*. Nairobi: East African Educational Publishers.

Okia, O. 2004. In the interests of community: Archdeacon Walter Owen and the issue of communal labour in colonial Kenya, 1921–30. *Journal of Imperial and Commonwealth History*, 32.

Owen, V. 1990. *Leonard and Gladys Beecher – A Life of Service to God in East Africa*. Privately published.

Parsons, W. 1934. *Travels of Mrs Judy O'Hea and Mrs Winifred Parsons 'Pub Crawling Through Africa'* – Kenya and Uganda. Bodleian Library/Rhodes House, MSS.Afr.s.1797.

Pavitt, N. 2008. *Kenya – A Country in the Making 1880–1940*. New York: W. W. Norton & Co.

Perham, M. 1963. *The Colonial Reckoning – The Reith Lectures*. London: Collins/Fontana.

Perham, M. 1974. *East African Journey*. London: Faber & Faber Ltd.

Pringle, J. W. 1893. Slave trade in connection to the projected railway, Ch. VIII. In *Mombasa Victoria Lake Railway Survey*. London: HMSO.

Richards, S. 1988. The South Eastern Agricultural College and public support for technical education, 1894–1914. *Agricultural History Review*, 36, 172–87.

Robertson, C. C. 1997. *Trouble Showed the Way: Women, Men and Trade in the Nairobi Area 1890–1990*. Indiana, US: Indiana University Press.

Ross, W. M. 1968. *Kenya from within – A Short Political History*. London: Frank Cass & Co.

Somjee, Dr. Sultan H. (Curator) 2000. The Asian African Heritage Exhibition.

Spencer, I. 1980. Settler dominance, agricultural production and the second world war in Kenya. *Journal of African History*, 21, 497–514.

St Olaves School. 1916 to 1970. *The Olavian* (School magazine).

Storrs-Fox, D. and B. 1920 to 1942. Letters to parents. Bodleian Library/Rhodes House, MSS.Afr.s.1029/A296.

Sunderland, D. 2004. *Managing the British Empire – The Crown Agents 1833–1914*. London: The Royal Historical Society.

Sunderland, D. 2011. *Economic Development of Africa 1880–1939*. London: Pickering & Chatto.

Swann, A. 1982. ODRP papers. Bodleian Library/Rhodes House, MSS. Afr.s.1717(149a).

Swynnerton, R. J. M. 1970. Commonwealth Development Corporation. Nucleus: Estate and smallholder programmes. In *Change in Agriculture,* ed. A. H. Bunting. University of Reading. London: Duckworth.

Thurston, A. 1987. *Smallholder Agriculture in Colonial Kenya: The Official Mind and the Swynnerton Plan*. Cambridge African Monographs 8. Cambridge: African Studies Centre.

Trench, C. C. 1993. *Men Who Ruled Kenya: The Kenya Administration, 1892–1963*. London: Radcliffe Press.

Turbo Kipkarren Association, 1928 to 1940. Records of the association. Bodleian Library/Rhodes House, MSS.Afr.s.881.

Uvarov, B. P. Anti-Locust Research Centre. 1951. *Locust Research and Control 1929–1950*. London: HMSO.

Veasey, R. 1925–27. Letters to his wife. Bodleian Library/Rhodes House, MSS.Afr.s.1970.

Wainwright, R. E. 1979. Responses for the ODRM. Bodleian Library/ Rhodes House, MSS.Afr.s.1717/161.

Ward, Kevin and Brian Stanley. 2000. *The Church Mission Society and World Christianity, 1799–1999*. Richmond, UK: Curzon Press Ltd.

Watkins, E. 1995. *Oscar from Africa: A Biography of Oscar Ferris Watkins, 1877–1943*. London: Radcliffe Press.

Watkins, E. 1997. *Jomo's Jailor*. Watlington, UK: Britwell Books.

Watkins, E. 2005. *Olga in Kenya – Repressing the Irrepressible*. Brighton, UK: Pen Press Publishers Ltd.

Watt, J. A. S. 1895 onwards. Bodleian Library/Rhodes House, MSS. Afr.s.391.

Wills, C. 1953. *Who Killed Kenya?* London: Dobson.

Wilson, F. Memoirs; Ch. 15, District Agricultural officer, North Nyanza, 1948–1950. (Unpublished memoirs)

Wood, S. 1964. *A Fly in Amber*. London: Harvill Press Ltd.

Wrong, M. 2009. *It's Our Turn to Eat: The Story of a Kenyan Whistle Blower*. London: Fourth Estate.

GLOSSARY OF TERMS AND ACRONYMS

ACK	Anglican Church of Kenya
ADC	Assistant District Commissioner
AGR	Department of Agriculture in Kenya
ALDEV	African Land Development Board
AO	Agricultural Officer, sometimes known as *Agricola*
ARP	Air Raid Post
BEADOC	British East African Disabled Officers' Colony
C in C	Commander in Chief
CMS	Church Missionary Society
DC	District Commissioner
DFID	Department for International Development
DO	District Officer
EAWL	East Africa Women's League
EU	European Union
Exco	Executive Council
FIDES	Fonds d'Investissement Économique et Social des Territoires d'Outre-mer
GDP	Gross Domestic Product
HMOCS	Her Majesty's Overseas Civil Service
IBEA	Imperial British East Africa Company
KAR	King's African Rifles
KCA	Kikuyu Central Association
KDF	Kenya Defence Force
KME	Kericho Music Ensemble
KWEO	Kenya Women's Emergency Organisation
Legco	Legislative Council
LNC	Local Native Council
MO	Medical Officer
NMS	Nairobi Music Society

ODA	Overseas Development Administration
ODM	Ministry of Overseas Development
ODRP	Oxford Development Review Project
OECD	Organisation for Economic Cooperation and Development
OWM	Old Wye Man
PC	Provincial Commissioner
PT	Physical Training
PWD	Public Works Department
RSCM	Royal Society of Church Music

Swahili terms

Banda	Hut, usually thatched with palm leaves
Baraza	Council meeting
Boma	Originally an enclosure for cattle made of hedges and thorn bushes; came to mean the government compound with offices and housing for government employees
Bwana	An important person, the boss; used as a form of respectful address
Duka	General store (often run by Indians)
Githaka	Land owned by Kikuyu clan or landowner
Meera	A leafy and quite addictive stimulant, often known as *khat*
Posho	Maize meal
Shamba	Field, small farm
Shauri	A disagreement, a to-do, a problem
Shenzi	Broken down, shabby, dirty, savage, untidy – a useful word
Sufurias	Flat based, deep sided, lipped and handleless cooking pots or containers
Wazungo	Foreigners
Wimbi	Millet

INDEX

Plate 1 Owen on safari, 1931.

Plate 2 Ridge tents on Mount Kenya, 1931.

Plate 3 Seed distribution at Embu, 1930.

Plate 4 Loaded for safari, 1930.

Plate 5 Owen at home in Nairobi 1932, with Boris.

Plate 6 Owen as Pish Tush, *The Mikado*, Nairobi 1933.

Plate 7 Owen and Dorothy arrive at Kilindini, 1934.

Plate 8 Arrival at their first home, Kakamega, 1934.

Plate 9 Getting around could be tricky, 1934.

Plate 10 Their home in Kisumu, 1935 – still standing and inhabited in 2010!

Plate 11 Malakisi camp, Dorothy in the centre calling Boris. The agricultural instructor and local chief stand to the right, 1939.

Plate 12 Department of Agriculture, Kisii, 1939.

Plate 13 New calves born on experimental farm, Kisii, 1939.

Plate 14 Washaway on Kano plains, 1930s.

Plate 15 Nairobi Cathedral Choir, 1948. Owen is in the back row, extreme left. Rosemary is half concealing him. Rev Hugh Evan Hopkins is in the centre and Cicely Barton on his left.

www.ingramcontent.com/pod-product-compliance
Ingram Content Group UK Ltd.
Pitfield, Milton Keynes, MK11 3LW, UK
UKHW020905190926
475560UK00010B/189

9 781780 769967